Norman Barichello

Gyrfalcon

The One Who Stays All Winter

Suite 300 - 990 Fort St
Victoria, BC, V8V 3K2
Canada

www.friesenpress.com

First Edition — 2019

Front cover photograph by Norman Barichello
Back cover photograph by Bryce Robinson

ISBN
978-1-5255-5102-4 (Hardcover)
978-1-5255-5103-1 (Paperback)
978-1-5255-5104-8 (eBook)

1. Science, Life Sciences, Zoology, Ornithology

Distributed to the trade by The Ingram Book Company

CONTENTS

I would like to dedicate this book to my wife, Barbara, whom I met in gyrfalcon country, and who encouraged me and patiently supported me throughout this book project, and to my son and daughter, Joshua and Rebecca, who have shared many of my encounters with gyrfalcons and my passion for the outdoors.

ACKNOWLEDGEMENTS

AS WITH MOST PROJECTS, I COULD NOT have done this without the support, advice, and inspiration of many individuals. Thanks first to Fred Bunnell at the University of British Columbia, who many years ago offered me the chance to study gyrfalcons, and Mike Gillingham, who generously helped me with data analysis. Dave Mossop was always willing to share his knowledge of ptarmigan with me, and we have shared many adventures with gyrfalcons. I want to thank especially Tom Cade, who encouraged me in no small way. He persuaded me to publish this book, he was most helpful in its review and design, and he graciously agreed to write the foreword. Unfortunately, Tom passed away in January before he could see this publication come to fruition and before I was able to meet him personally. I was inspired by three other extraordinary ecologists: Ian Newton, Bob Frisch, and Dena elder Charlie Dick. Danny Nowlan and Bob Collins helped me to understand more about captive gyrfalcons and falconry. Thanks also to Steve Van Zant, Peter Devers, Kent Carnie, and James Weaver, who took an interest in this story and provided helpful advice. I am very appreciative to Paul Webster, who was instrumental in the layout and design of this book, and to David Anderson, Jannik Schou and Bryce Robinson, who generously contributed their stunning photographs. Bryce, along with Michael Henderson and the Peregrine Fund, also kindly allowed me to use some of the images taken by remote cameras as part of Bryce's and Michael's graduate work. Also, thanks to Vadim Gorbatov for his generosity in sharing his extraordinary artwork, and to Alan Gates, who enabled me to exchange emails with Vadim. Inuit artist Abraham Anghik Ruben and Rocco Paneese of the Kipling Gallery also kindly shared images of one of Abraham's exceptional sculptures. Alexander Stubbing and Edward Atkinson of the Nunavut Heritage Centre, with the blessing of the Nunavut Government, graciously provided me with images of a remarkable ancient Dorset carving. Thanks also to Gurinder Mann, who brought to my attention a painting of Gobind Singh, and to the Victoria and Albert Museum, which were willing to share this image with me. I am grateful to the World Wildlife Fund, which provided the financial assistance that allowed me to introduce myself to gyrfalcons in the central Yukon.

FOREWORD

BY TOM J. CADE

PROBABLY NO BIRD HAS BEEN MORE ADMIRED and sought after since prehistoric times than the magnificent gyrfalcon, living illusively in some of the remotest and difficult landscapes of the circumboreal subarctic and arctic regions of the world. Since the earliest records about human activities, it has been prized by falconers for use in hunting because of its large size, hunting skills, and great beauty, especially the white ones. Only the more common, universal peregrine might challenge the gyr for desirability as a hunter.

As one might expect of such a popular bird, much has been written about the gyrfalcon in the scientific, historical, and popular literature. Popular books have greatly increased in recent years, ranging from children's storybooks to mystery novels, and even a book of gyrfalcon poems (Collin Simms, 2007, Shearsman Books). It is appropriate to mention some of the more outstanding examples as background for Norman Barichello's book. A quick search of the internet reveals dozens of books and monographs in several languages, not including the more technical scientific reports.

When I first began my studies on gyrfalcons in the 1950s, there were only a few important sources of information available on these birds. By far the most significant one was the 1943 translation from Latin to English by Casey A. Wood and F. Marjorie Fyfe of the thirteenth century treatise written by the Emperor Frederick II of Hohenstaufen and his son, King Manfred, *De Arte Venandi cum Avibus*, a tome that includes thirty-two "chapters" about the natural history, behavior, and training of gyrfalcons. Other books and sources at that time include natural history accounts such as Arthur C. Bent's *Life Histories of North American Birds of Prey* (1938), Georgy Dementiev's Russian book, *The Birds of the Soviet Union, Vol. 1*, on birds of prey. This was later translated into English, and his *Sokola Kretcheti* (1951) and a German revision was published in 1960 on the life history and human relationships. It was also later translated into English. Yngvar Hagen's 1952 report on Norwegian gyrfalcons called attention to the central importance of ptarmigan in the diet, and there were some adventure stories such as the one by the one-eyed, one-armed Ernest Lewis [pseudonym for E. Vesey], who traveled to northwest Iceland to capture young gyrs for falconry. It was published in 1938. A similar trip to Iceland was reported by Stanley Cerely (1955). Other references available at the time can be found in my 1960 publication.

Since the publication of my thesis in 1960 on the ecology of gyrfalcons and peregrines in Alaska, there has been a steady growth in both the scientific and popular literature about gyrfalcons. Much of it has come since the turn of the century and focused on the impending problems predicted by the rapid rate of climate warming occurring in subarctic and arctic regions. Perhaps most noteworthy has been the long-range scale of work carried out in Iceland by Olafur Nielsen (Nielsen and Cade 2017), now approaching fifty years of continual research on both gyrfalcons and ptarmigan. Additionally, there have been long-term studies carried out in Greenland by William Burnham, Curt Burnham, Bill Mattox, Travis Booms, Mark Fuller, and others, as well as in Scandinavia, Finland, and Russia.

Much of this work has been summarized in various reviews and in one major book simply titled *The Gyrfalcon* (2005), in which Eugene Potapov and Richard Sale summarize most of the world's literature on the gyr up to that date. Other important sources include the Birds of North America Accounts sponsored by the Cornell Laboratory of Ornithology (https://birdsna.org) [see the most recent gyr account by Travis Booms, Tom Cade, and Nancy Clum]; *Birds of the Western Palearctic* update 2(1):1-25, 1998, by Cade, Pertti Koskimies, and O. K. Nielsen; *Gyrfalcons and Ptarmigan in a Changing World*, two volumes (2011), edited by R. Watson, T. Cade, M. Fuller, G. Hunt, and E. Potapov; A*pplied Raptor Ecology, Essentials from Gyrfalcon Research* (2017) edited by D. Anderson, C. J. McCllure, and A. Franke and published by The Peregrine Fund. A popular and well-illustrated book combining falconry and natural history by Emma Ford (1999) should also be mentioned.

Now comes this new book by Norman Barichello, who has spent most of his life in the Yukon Territory of northwestern Canada as an outdoorsman devoted to a wide variety of activities associated with management and conservation of mammalian predators and big game animals. He received a master's degree from the University of British Columbia in 1983 after five years of intensive fieldwork, summer and winter, in the Ogilvie Mountains studying the ecology and behavior of gyrfalcons and their associated ptarmigan prey. Since 1991, he has been the co-owner and manager of an ecotourism business in northern Canada, providing him with additional opportunities to observe his two favorite birds. As revealed by the title of his book, Norman has also been strongly involved in efforts to preserve and understand the traditional knowledge of the aboriginal people in northwestern Canada.

In addition to describing many other interesting aspects of the gyrfalcon's natural history in arctic Canada, Norman has also provided the reader with a detailed and updated account of the history of the gyrfalcon's many relationships with humanity, from the earliest prehistorical suggestions and the earliest writings in China, to the rich literature of the Middle Ages and Renaissance. These covered the commerce and trade in gyrfalcons, so extensively used for falconry universally by the world's aristocracy. He also discusses the current captive breeding centers that have proliferated worldwide since the first successful breeding attempts by The Peregrine Fund at Cornell University in the 1970s. Norman's scholarly chapter on human/falcon relationships now largely supplants the previous fine essay by Georgy Dementiev (1960) on the same subject.

True devotees of the Gyrfalcon will find it difficult to put down this book until it has been read from cover to cover.

BEGINNINGS

AS DAWN WAS BREAKING, THE MALE GYRFALCON catapulted himself from his favorite perch and into his vast hunting domain. He was hungry and alert. It was another magnificent day. A light dusting of snow gave the land an almost spiritual aura, with the mountains standing as monasteries in this clear cold air.

At first he flew along the ridges, which offered him slight updrafts that aided his flight and afforded him a broader view of the subarctic world. In the distance was a series of mountain ranges, interrupted by the odd peak that penetrated the heavens. Below him was a land devoid of forests, where the monotony of snow was broken by the occasional rock outcrop and willow-lined creek draw.

Experience from previous hunting drew the male gyr to these basins and draws where the willows extended above the snowline. It was here that the ptarmigan were apt to be found. They were white in a snowbound world and in flocks, seeking food from the willow buds and refuge from gyrfalcons. He dropped down and began a low-level search for his favorite prey.

It didn't take the gyr long to notice movement in a willow draw as he came around a bend. If he could slip in undetected, he might force one or more of the ptarmigan into the air, to his advantage. He focused on a small isolated patch of willows and began his attack, low to ground, using the terrain to hide his approach and his momentum to minimize the movement of his wing beat.

He was almost there. Suddenly the ptarmigan scattered in a blur of white. Three managed to get behind him and were gone. Another half-dozen scrambled into the thick willows and froze. The rest were able to get a slight jump on him and disappeared into the willows on the opposite side of the draw. The gyr flared and circled over the willows, more out of curiosity than with any realistic prospect of a second chance. His advantage was surprise, and this time he was unable to catch any of them off-guard.

He continued his search for another hour. Two more chases, but neither was successful. Ahead was a mottled black rock splashed with bright orange lichens along a ridge of fractured outcrops. The spot offered a vantage point over an area of broken topography with a number of small creek draws that fed into a main artery. He discretely settled on the orange-splashed rock.

It was perhaps a half-hour later when the moment presented itself. A flock of a dozen or so ptarmigan scuttled around the corner and out of the dense willows, feeding as they went, oblivious to the impending attack. The gyrfalcon launched quickly from his perch taking advantage of the moment. His flight was true and fast, and again low to the ground. On the backside of the hill, he was immediately among them. Startled and with no time to find satisfactory cover, they exploded into the air. The gyr was ready for them and made no mistake. He struck one of the ptarmigan with an extended rear toe and with a puff of feathers it fell to the ground. The gyr immediately dropped on it, pinned it, and grabbed its neck with his beak. Inserting the little teeth on his upper mandible between two vertebrae and with a deliberate twist, he dispatched his prey and the hunt was over.

He tore into the carcass to taste the fruits of his victory and replenish his energy but reserved much of his catch for his mate. The trip home was laborious but invigorating. Miles away she saw him, no doubt aware that his compromised flight meant food, and she flew to him. As she approached, he gained his second wind and teased her with his catch. They danced momentarily in the skies above the nest cliff, then he came to her and gave her his prize in the air. Calls filled the amphitheater created by the cliffs, further electrifying the mood.

The male gyr perched nearby and watched as she ate. He was a handsome fellow with his light grey back, contrasting a white breast, subtly barred with black arrow-shaped etchings. Bright yellow legs, visible below feathered leggings, drew attention to his large imposing talons. His coal black eyes were accentuated by a bright yellow eye ring and a dark shadow extending onto his cheek, which gave him a regal look.

Part way through her meal, she paused and began making subtle seductive gestures. Caught up in the ecstasy of courtship, he flew to the nest where she quickly followed. They alit on the edge of the nest and faced one another with heads bowed and vocalized to one another, both aroused by the excitement of bringing youngsters into the world.

INTRODUCTION

BRINGING THE TUNDRA ALIVE

It was one of those remarkable winter days in the Yukon's Ogilvie Mountains in northern Canada. It was -45^{0} and calm. The treeless subarctic tundra was draped in fresh snow, creating a mosaic of whites and blues, the effect of long shadows cast across broken terrain. The willows, rising above the tundra and giving the scene a brush of texture, were covered in crystal flakes of ice like diamond tiaras. In the distance was a series of rugged mountains set against a cerulean sky. It was the epitome of subarctic winter—cold, pure, desolate, and strikingly beautiful.

Suddenly the serenity was broken and the tundra was brought to life. A flock of twenty or so ptarmigan in their white winter plumage exploded out of the willows, scattering the ice crystals into a kaleidoscope of light. Above them were two large grey-white falcons, one in front, one following, piercing the sky at about 160 km/hour just above the ground. Then they were gone, and the land was again silent.

This was the beginning. I was to go on to study gyrfalcons in the central Yukon for six years, but was to be a casual student of these large falcons for many more years. I explored the relationship between gyrfalcons and ptarmigan and listened to "old-peoples'" stories about these magnificent falcons. I also tried to understand the passion for falconry over the millennia, and I continued to puzzle over those aspects of the gyrfalcons' life history that eluded me.

As a wilderness guide for many years, I have met many naturalists hungry for stories about nature and particularly curious about those iconic species that are less common. Other than general descriptions about wildlife, or identification guidebooks, most narratives about gyrfalcons are told through scientific papers and manuscripts. Such a medium is typically specialized and technical, uses terminology best understood by scientists, and ignores observations that are difficult to test. In writing this book about gyrfalcons, it was my hope to tell a story that will appeal to naturalists and falconers who may have little formal scientific training but nonetheless curious about the ecology of these extraordinary birds and their remarkable association with humans.

THE ECOLOGICAL CHALLENGE

Living in the far north in the winter is indeed a challenge. Daylight is reduced to a few hours, temperatures are severe, and much of the land is blanketed by snow. Of no surprise, there is little evidence of life. Many northern creatures escape the rigors of cold and insufficient food by hibernating or migrating. Of about eighty species of birds that nest in the area, only eight overwinter. Most of the small mammals are below an insulating cover of snow, either sleeping in dens or concealed by the snow cover.

Yet, here lives the gyrfalcon. Not only challenged by brutally cold temperatures and few available daylight hours to hunt, gyrs are also constrained by the availability of prey. Ptarmigan are the only real choice, the only resident that is accessible above the snow (at least for a portion of the day), is of a size that warrants a chase, and occurs in numbers that could provide a sustainable diet. But catching them is by no means assured.

Winter scene in the Snake River area, northern Yukon

Ptarmigan are superb fliers. They have inherited relatively short broad wings and heavily developed breast muscles that allow them to outfly and outmaneuver most predators. They also live in flocks that give them the advantage of many eyes to detect danger. Ptarmigan also change their plumage twice a year as camouflage against predators. Their strategy is simple: avoid being seen and never go too far from the willows that provide cover from their principal predator, the gyrfalcon. Ptarmigan have also adapted to their winter environment by spending considerable time in snow cavities. In the bowels of the snow, they enjoy the advantage of a relatively moderate temperature and no wind, and perhaps more importantly, they are hidden from gyrfalcons.

The gyrfalcon is further handicapped because their nesting period is roughly five months in an environment where summer lasts a mere three. To have any chance of success, gyrs must begin nesting early, while winter prevails. At a time when just surviving imposes a significant challenge, the gyrfalcon initiates a demanding reproductive schedule.

As if cold, limited daylight, and dependence on a single elusive species that frequently avoids attention by snow roosting isn't enough, the gyrfalcon faces a further challenge. Ptarmigan populations are cyclic. For a few years, they are abundant, and then they suddenly crash. Instability is always a hardship, but in the north where the environment is simple and severe and where choices are meager, it can be perilous. Not only does this cause year-to-year uncertainty, it creates a boom and bust phenomenon that has significant demographic implications for gyrfalcons.

But before I try to explain some of these ecological relationships, I will do my best to describe the gyrfalcon, including its appearance and its adaptive traits as shaped by its ecological setting.

Young gyrfalcon overlooking the Snake River valley in the northern Yukon

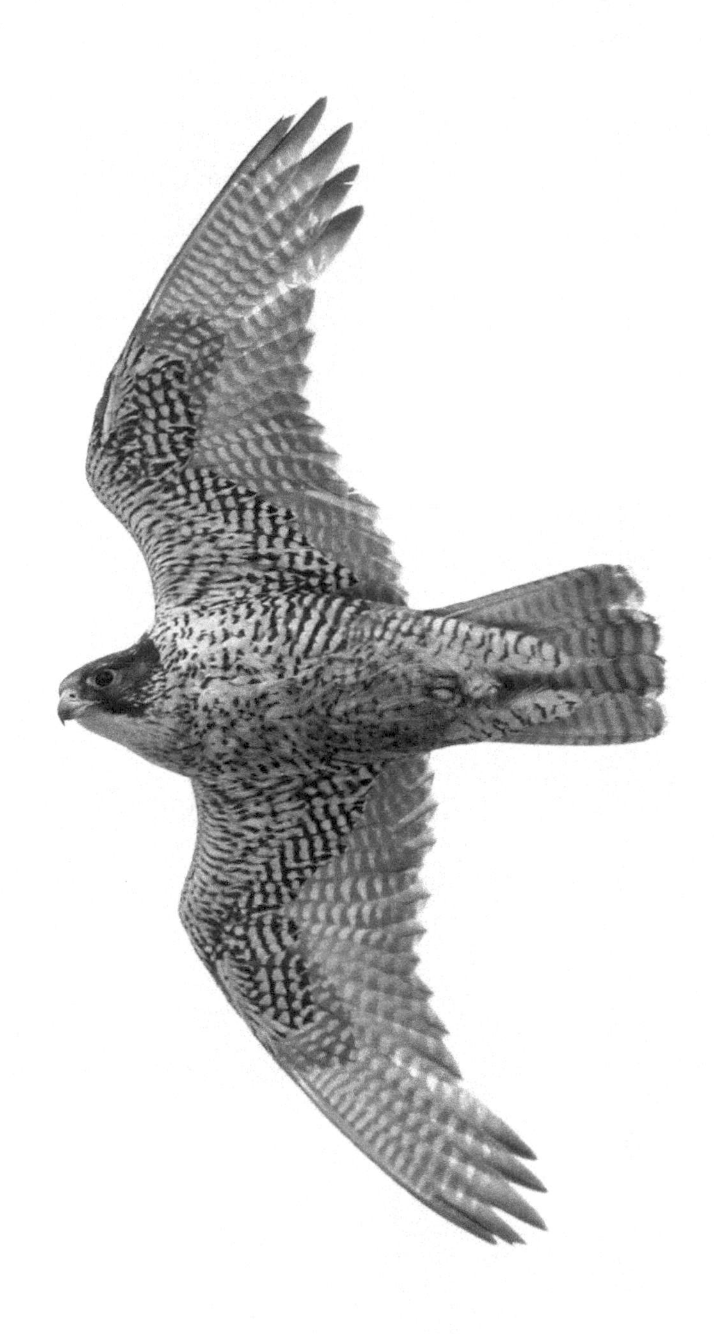

CHAPTER ONE

AVIAN PERFECTION

FEW ARE BLESSED WITH THE OPPORTUNITY TO see a wild gyrfalcon. Gyrs occur at low densities in the north and only a few venture south. They live in the most isolated parts of the planet, discouraging most of us from seeking their secrets. Here the gyrfalcon is easily identified: a large elegant bird of prey, weighing between 800 to 2100 grams (or 2.6 – 4.6 pounds) and about the size of a large raven. I was to appreciate their formidable stature on my first descent into a gyrfalcon eyrie to put leg bands on the youngsters (to learn more about where they spend the winter). Suspended forty meters above the ground, clutching a climbing rope, still learning the knack of rappelling, I glanced up to find myself at the mercy of an adult gyrfalcon. She was accelerating toward me at blistering speed, all 2000 grams of her, with talons outstretched. Her guttural screams, amplified by the amphitheater of cliffs that surrounded her nest, made the experience even more terrifying.

Taxonomists have found a home for the gyrfalcon within a group of birds called falcons, all of which are part of a larger family known as Falconidae. Ecologically, falcons are a group of hunters that typically reside and hunt in open landscapes, enabled by relatively long-pointed wings capable of high speed. They have talons that give them the ability to hold and grasp their prey, and hooked beaks to enable them to tear the carcass into digestible chunks. These features are what define them as birds of prey, commonly referred to as raptors, derived from the Latin word rapt, which means "to plunder or seize." Within the falcons, gyrfalcons are lumped into a sub-group called the Hierofalcons (Wink et al. 2004). In addition to the gyrfalcon, this group includes saker, laggar, and lanner falcons. They all have patterned plumage and their flight is typically level when in pursuit of quarry (although gyrfalcons, sakers, and lanners are capable of high-speed diving, which is the hallmark of peregrine falcons).

Unique to the falcons, as well as kites and a few accipiters (hawks), is the occurrence of a small, sharp pointed protrusion on each side of the upper part of the beak that biologists call the "tomial tooth." This attribute gives falcons the ability to kill prey quickly by biting into the neck to dispatch their prey and

Female gyrfalcon

avoid a risky struggle. Most falcons are capable of killing airborne prey by striking at high speed made possible by long pointed wings. But speed and agility come at the cost of having a less robust frame. Struggling with prey runs a high risk of injury so the efficacy of the "tooth," which can penetrate the vertebrae and dispatch prey quickly and efficiently, has great utility. As Peter Dunne described, it is "like a hand held can-opener, designed to open the cervical vertebrae of prey, letting the life out" (Dunne 1995).

Talons tell us a lot about a predator. They are an essential weapon to a raptor, tailored to the size and disposition of their primary prey. The gyrfalcon's feet are large, but not quite as imposing as those of a goshawk or a great-horned owl, which similarly choose large prey. Gyrfalcon talons are designed to disable relatively large prey with a mid-air strike with toes spread wide apart. Their talons are large enough to provide blunt force and to grab and handle large prey, but not as threatening as those of the goshawk or great-horned owl, which lack the tomial tooth to kill prey outright, and use their talons like vice-grips to kill by squeezing.

The gyrfalcon's airfoil is also precision designed. The wings are long and pointed but not as exaggerated as that of the peregrine, its smaller and well-known cousin. This heavier "wing-loading" enables the gyr to sustain speed over a longer distance and maintain speed when climbing. It is well known among falconers that gyrfalcons on the attack can climb and maintain speed at a steeper angle than most birds. Compared to other falcons, their range of attack is said to be two to three times greater (Nelson 1956). The gyr also has a longer tail than most falcons, serving as a rudder. The superb engineering of the falconidae then is further adapted in the gyrfalcon for sustained speed and agility. This airframe makes the gyr one of the most formidable avian predators, capable of hunting both birds and mammals, large and small. Emperor Frederick II of Hohenstaufen, in his thirteenth century treatise on falconry, *De Arte Venandi cum Avibus*, wrote of the gyrfalcon, "Out of respect for their size, strength, audacity and swiftness, the gerfalcons shall be given first place in our treatise," and, she "holds pride of place over even the peregrine in strength, speed, courage, and indifference to stormy weather" (Wood and Fyfe 1943).

Portrait of a white gyrfalcon. Note the large eyes and tomial tooth

Most descriptions of gyrfalcons note their exceptional eyesight, a trait well known to falconers. Birds—and birds of prey in particular—have a number of specialized features that provide them with exceptional vision (see Walsh and Milner 2011). There are typically two main types of visual (photoreceptor) cells in the retina of the eye. There are the cone cells that provide color, and there are the rod cells, which are much larger and extremely sensitive to light. These rod cells enable vision under low light conditions and improve motion detection. The proportion of rods and cones reflect lifestyle. Diurnal birds have more cones than rods in their eyes (approximately 80% of the photoreceptors are cones). Owls, by contrast, are largely nocturnal hunters and have few cone cells (Sankaran 2012). Due to the larger physical size of rod cells as compared to cone cells, the eyes of nighttime hunters are typically much larger in proportion to body size than are daytime hunters. In fact, the eyes of some owls exceed the size of our own.

This ability of some birds of prey to see under low light is an added bonus to an otherwise remarkable eye, says Stephanie Streeter (2001) in a piece called "The Eyes Have It". The density of visual cells in their eyes is many times greater than that of humans—in the neighborhood of one million visual cells per square millimeter as compared to our meager 200,000 cells per square millimeter. In addition to this, falcons possess an extraordinary "power of accommodation" of the eye achieved by their ability to adjust the shape of the cornea and the lens to allow rapid change in focusing (Jones et al. 2007, cited in Walsh and Milner 2011). Birds of prey also have a retina that is twice as thick as a human's and have a much greater number of ganglion cells than our own and so can send more visual messages to the brain (Sankaran 2012). The eyes of birds of prey are also set toward the front of the face, allowing the field of visions to overlap. This gives most raptors excellent depth perception to allow them to triangulate their field of vision.

Birds also benefit from the ability to process a high frequency of images per second, called Flicker Fusion Frequency. This refers to the frequency at which intermittent flashes of light appear to be completely steady—in other words, the rate at which the images from individual frames become blended. This is what makes movies possible: by streaming separate images at a certain frequency so that the moving image is smooth and continuous. With increasing Flicker Fusion Rates, more frames can be interpreted. The continuous stream of vision is achieved at about twenty-four frames per second in humans, and about 100-120 frames per second in birds (Lederer 2016). This means that avian predators are able to differentiate at least four times more frames of vision than humans are. It would be akin to watching events in slow motion. Tom Cullen, a raptor breeder and lecturer, provides a compelling example of how this works. In baseball, the distance between the mound and home plate is sixty feet. At twenty-four frames per second, the batter has about four frames of vision to predict the location of the ball. Supposing a gyr receives four times more frames of vision, the bird would have a much greater ability to resolve visual detail, and so achieve a much more precise prediction of where the ball is over the course of its flight. In practical terms, a high Flicker Fusion rate provides gyrfalcon with more capability in interpreting the flight of a ptarmigan and responding appropriately.

Although to my knowledge the morphology of the gyrfalcon eye has not been scientifically studied, we can infer by their relatively large size, the bird's obvious ability to see under low light conditions, and the remarkable features of avian eyes—and birds of prey generally—that the gyrfalcon's eye holds proportionally more rod cells than other daytime hunters. During the winter in the north, the days are short and the available light for a daytime hunter is limited. The ability to see during those extended dawn-dusk periods may not only be an advantage, but a requirement for this northern hunter—"the one who stays all winter."

Gyrs are also sensibly dressed for the severe weather. They have dense down feathers that effectively trap air giving more loft and buoyancy and therefore more insulation than is typical in most birds. Their feathers extend partially down the legs in the form of leggings similar to that of other northerners—the ptarmigan and the rough-legged hawk.

Female gyrfalcon on her nest clutching an arctic ground squirrel

Large size is an added advantage in cold climates. Most northern animals are larger than their southern counterparts. Note the moose, bison, polar and grizzly bears, arctic and subarctic wolves, and even arctic hare. Increasing size improves the ratio of body size to surface area, and therefore reduces the amount of heat loss. The gyr too has the advantage of relatively large size.

Size is also associated with the rate at which calories are burned. Most small birds with high metabolic rates cannot go without food for long, but the gyrfalcon with its comparably large size and its ability to store fat can go without food for long periods, allowing them to persist in a land with few prey, scattered widely (Cade 2011). It also permits the gyrfalcon to wait out severe storms in sheltered alcoves and crevice perches.

The gyrfalcon has other ways of reducing heat loss. According to well-known falconer Frank Beebe (1976), the gyr has a unique ability to rotate its hind toe inward so that it points parallel with the other toes. This enables the bird to lie on its chest with the legs tucked up underneath it and hunker down to reduce airflow and protect its extremities and so minimize heat loss. Beebe claims that gyrs can go to sleep in this position. Tom Cade (2011) also remarked, based on observations of lethargic behavior of captive gyrfalcons, that gyrs may be able to drop their body temperature a few degrees to reduce metabolic rates when they cannot hunt for food. Cade suggested that even a modest drop in body temperature can reduce the energy requirement for maintenance to half the normal rate. He and Olafur Nielsen surmised that with a slight reduction in its metabolic rate and body temperature a 1,500 gram gyrfalcon with 150 grams of fat should be able to survive for 10-14 days without eating, and still be capable of hunting (Nielsen and Cade 1990). Bob Collins, a director of The Peregrine Fund and ardent falconer, also remarked that his captive falcons would often plunge into soft snow and roost with only the head exposed. Could this also be a behavioral adaptation, similarly used by ptarmigan, to minimize heat loss?

The physical attributes of gyrfalcons go hand in hand with their habitat requirements that include alcoves or crevices on the cliffs of their home base. I have visited over seventy-five gyrfalcon nest sites across northern Canada. All have included at least one crevice perch, as evident by the heavy deposit of excrement exuding from these holes in the cliff. Cade (1960) noted similar perching sites on cliffs in Alaska's Arctic. For the gyr, the cost of hunting in severely cold weather, combined with few daylight hours and periodic storms, may at times outweigh the benefits of a long hunt. Also, when the weather is severe, ptarmigan are more apt to be in snow roosts hidden from gyrfalcons. Perhaps the gyrfalcon has adapted to these circumstances by entering crevices in severe weather where they are able to hunker down in a prone position and induce a state of hypothermia, thereby reducing body temperature and metabolism, and the extreme energetic drain required for hunting.

Gyrfalcon attire is both striking and functional. Their plumage comes in three predominant colors. The most common are grey and white with varying degrees of contrasting bars, stripes, spots, and shades, including silver-colored gyrs (Johnson and Burnham 2012). White gyrs can be pure white or silver, or flecked with black or grey. The grey gyrs can be light grey to slate grey and even blue grey or with brownish tints, again with variable dark markings.

But there are other less common color variants. One bird I observed on Baffin Island in northeastern Canada was a light tawny color. Tom Cade also observed a tawny colored gyr in northern Alaska, as well as a bronze colored gyrfalcon wintering south of Canada. Another color variant in the central Yukon was a "ladder-backed" form with contrasting black, white, and grey markings. There are also spotted gyrfalcons. A well-known Russian ornithologist, Dr. Georgiy P. Dementiev, in his 1960 German treatise on gyrfalcons, mentions spotted gyrfalcons as "somewhat unfortunate deviations from the birds of the usual coloration," and refers to the list of hawking birds of the Tsar as including seven spotted and three black, from a total of ninety-nine gyrfalcons. There is also mention of a piebald gyrfalcon, possibly a spotted gyrfalcon, which was sent by a Persian ruler to a Mughal emperor in India in 1619 (Allsen 2006).

The occurrence of different colors varies across their range. Among the three most common colors, black gyrs are the least common and occur in pockets primarily in northern Labrador and Quebec, southern Greenland (at least formerly), and perhaps the Altai Mountains (T. Cade, pers. comm.). In my searches across Canada for gyrfalcons (on Baffin Island, in the central Arctic, and in the north and central Yukon), I have found varying mixes of white and grey gyrfalcons. On Baffin, about 80% of the gyrs I saw were white. In the central Canadian Arctic, about 50% were white (a similar percentage reported by Poole and Bromley (1988)). I have never seen a white gyr in the Yukon. Yukon ornithologist Dave Mossop claims they are rare in western Canada, although in parts of northwestern Alaska and even on the North Slope in foothills of the Brooks Range white gyrs occur regularly at some nests (Tom Cade, pers. comm.).

White gyrs tend to be more common in areas that are snow bound for much of the year—typically in northern latitudes and notably in northeastern Canada, Greenland, and northeastern Siberia. "It's probably about ocean currents," says David Ellis and co-authors (1992). They argued that where warm ocean currents enter the Arctic (i.e. the Norway Current), the gyrfalcons are predominantly grey, and where cold currents surround areas of the Arctic (i.e. the Anadyr and Oya Currents), we find predominantly white gyrfalcons. Cold currents, no doubt, yield persistent snow cover.

Also, the distribution of these two predominant color variants mirrors that of two important antagonists. The distribution of grey gyrfalcons in North America generally overlaps with the breeding distribution of golden eagles. White gyrs, on the other hand, share their range with snowy owls. Both golden eagles and snowy owls pirate from gyrfalcons, probably killing those gyrs pre-occupied with their kill. Golden eagles in particular arouse such ferocity from gyrfalcons that seems to imply something far greater than simply competition for food or nests. The gyr's reaction to snowy owls is similar. Plumage color then is likely about camouflage; an adaptation to minimize fatal encounters with these large predatory pirates, as well as conceal them in the pursuit of prey.

Selective pressures may reinforce the advantage of different colors across their range. In central-west Greenland, where multiple colors are present, white gyrfalcon males were found to have earlier hatch dates than silver or grey males, whereas no differences were evident for females (Johnson and Burnham 2012). As laying date is probably indicative of productivity (as I discuss later) and males are the primary provider during the first half of the nesting period, these observations imply that white males have an advantage in environments with persistent snow.

Grey gyrfalcons display subtle variation in shades and markings. Juvenile birds are typically grey-brown. Adults tend to be greyer, with males tending to be a lighter grey than females. Within genders, there is also a gradient from light to dark grey. There are differences in breast patterns related to age and gender (Potopov and Sale 2005). The appearance it gives is one of horizontal barring in adult birds and vertical streaks on juvenile birds. In the Yukon, barring is most evident on female birds. Generally, feather etching on the breast becomes less pronounced as the gyr ages, with males having more subtle markings than females.

Noticeable on most, if not all, grey gyrfalcons is a black stripe that extends below the eye called a malar stripe. Not as pronounced as on the peregrine, but nonetheless noticeable. These stripes vary from being mere shadows to very prominent. Another well-known, high-speed predator, the cheetah, also sports black eye stripes. These "shadows" likely help reduce glare, an advantage not lost on American football quarterbacks and tennis players, who apply similar eye-shadow to help minimize glare and enhance depth perception.

Adult gyrfalcons are particularly striking. They display brilliant yellow markings on the bare skin around the eyes, beak, and on the legs. These yellow features are found only on adult birds and are most vivid during the courtship period. The accentuation of these yellow markings is well known to falconers in the business of captive breeding, said the late Danny Nolan, a gyrfalcon expert and enthusiast from the Yukon. Presumably hormonal changes combined with body condition influence the intensity of the yellow, perhaps as an accessory to attract mates.

Young birds do not have these yellow markings. Rather, they have blue-grey markings around the eyes, beak, and on the legs, and darker, heavily streaked plumage. Skuli Magnusson (1785), in describing the capture of gyrfalcons in Iceland (cited in Aegisson 2015), claimed, "The gyrfalcon's age can be fairly accurately estimated until the fourth year, for one and two-year old falcons have bluish white feet, in the third and fourth year their feet are a little yellowish, turning complete yellow from then on."

Young male gyrfalcon near fledging. Note blue-grey feet, cere and eye-ring

These distinct blue-grey markings possibly keep young gyrs safe from adults. Adult gyrfalcons of the same sex are aggressive towards one another within nesting areas and perhaps more generally. If young birds were unrecognized by adults, they could be easily killed or injured as potential competitors. By displaying neutral colors, young gyrs possibly avoid persecution during a time when they offer no competitive threat to adult birds. It may also help conceal them from golden eagles and other potential pirates and predators (it should be noted that these are hypotheses that have yet to be tested by definitive research).

The color variation between males and females also works to their advantage. Females spend much more of the time during the nesting period incubating eggs and protecting the nesting cliff from eagles as well as other nest predators. A darker grey with mottled breast provides her better camouflage against the cliff. The male's plumage is best suited for hunting, with a light and uniform underside to minimize contrast against light grey skies.

Most people associate large size with males—bigger, stronger, and more physically dominant—but this is not the rule for most birds of prey. Females are typically larger than males, sometimes half again as large as the male of their species. The gyrfalcon is no exception. Females weigh between 1,250 to 2,000 grams, while males weigh only about 1,000 to 1,400 grams (Cade 2011). The difference can be obvious. In flight, females tend to have a deeper wingbeat while males have a shallow crisp wingbeat. This gives the male more grace and agility when flying. When perched, the female gyr usually dwarfs her mate. The commonly used name for a male falcon, "tiercel," likely derives from the Latin word tertius, which means one third, as males are roughly a third smaller than females (Oggins 2004). Biologists call this reversed sexual size dimorphism; the female is larger than the male, which is the reverse of what we typically see in many mammals and some birds. The question is why.

Male and female gyrfalcon on their nest Note the significant size difference

In most mammals, size is linked to sexual advantage; bigger males out-compete their smaller rivals for breeding privileges. The larger individuals are more apt to win the physical contests or out-bluff the competition, assuring the female that they are more capable. Consequently, natural selection favors large males. The same may hold true for gyrfalcons (in reverse), where competition for mates and nest sites may be intense. According to Tom Cade, a recent ongoing study of peregrine falcons has found that intraspecific aggression over nest sites and/or mates (females against females, males against males) is a major cause of fatalities among adult peregrines.

Although this competitive pressure within genders may yield larger individuals, it does not adequately account for sexual dimorphism. Perhaps males and females are also sized for their respective roles during the breeding season. The male gyrfalcons may be the "right" size for hunting ptarmigan. A smaller frame and lighter weight gives him more agility and acceleration than the heavier female. This seems clear when watching the pair in aerial, acrobatic displays: the male with tighter turns at higher speeds.

One spring during courtship, I watched a male gyr, in attempting to land with a ptarmigan to present to the female at the nest ledge, accidentally drop his gift. It began a free-fall to the bottom of the cliff some fifty meters below. The female jumped off the nest to retrieve her prize, the male followed. He overtook her, grabbed the ptarmigan before it hit the ground, and broke out of the dive in a vertical ascent to deliver the carcass to the nest, where he waited for the female.

Female gyrfalcon

An apparent advantage is that being smaller, although a possible limitation to take down large prey, provides the necessary agility to hunt smaller prey, the ones toward the base of the food pyramid. This advantage opens up more prey opportunities. The male gyrfalcon, then, has a greater spectrum of prey choice than the female, without compromising his ability to hunt ptarmigan. A compelling diet study by Travis Booms and Mark Fuller (2003a) in Greenland, using motion-activated cameras, indeed found a diet shift to passerine birds later in the nesting period, most delivered by the male gyr. More recently, Bryce Robinson (2016) also recording prey deliveries through motion-activated cameras at gyrfalcon nests on the Seward Peninsula, found a shift as the season progressed toward medium and small prey.

The larger female may have given up some of the advantages of being a ptarmigan hunter in favor of being a mother and guardian during the nesting period. She bears and incubates the eggs, shelters and protects the young, and guards the nesting cliff. During the courtship period, the incubation period, and the first two weeks when the young are in the nest, she postpones her job as a predator and the male provides entirely for her and their family. Large size is a decided advantage around the nest. The larger female can go without food for longer periods—perhaps a prerequisite when she is sitting on eggs while the male is out hunting. She can also lay larger eggs, which is a distinct advantage in an arctic climate where a larger volume to area ratio minimizes heat loss. Large size also makes her a better incubator and brood hen. She has a larger brood patch to envelop the eggs and more size to provide shading when temperatures climb later in the nesting period. Large size is also an asset for nest defense. I need only remember my experiences at the end of a rappelling rope. Every close encounter I have had with gyrfalcons when I was too near the nest site was with a female.

Another advantage to large female size is to optimize hunting opportunities later in the nestling period at a time when the young are growing fast and growing feathers, and when the females leave the nest to hunt for the youngsters. The female is more capable of taking down large prey such as ducks and even geese and arctic hares. This ability has the decided advantage of expanding the prey base at a time when the male has depleted the supply of ptarmigan (White and Cade 1971). It would appear then that different duties are best served by different size—ergo, sexual size dimorphism.

A description of gyrfalcons is incomplete without saying something of their disposition. Falconers have many words of praise for the gyrfalcon in addition to their superb hunting ability. Frank Beebe (1976) claimed, although perhaps in a somewhat exaggerated opinion, that the gyrfalcon is by far the most responsive to human attention as compared to all other birds of prey. They are more relaxed and attentive than the single-minded and intense peregrine falcon. They begin training willingly and without hesitation, and appear to recognize their trainer by face. Some have described gyrs as playful (Cade 1953, Potapov and Sale 2005). My encounters with wild gyrfalcons draw me to the same conclusion: they are playful, accommodating, and perhaps curious.

Falconers claim that gyrfalcons are also temperamental. Every now and then, unpredictably, they become frantic (Stevens 1956). With this trait, they are not easily trained. It was the general opinion of English falconers that gyrfalcons were more obstinate and stubborn, but tolerant (Dementiev 1960). Consequently, accustoming gyrfalcons to hoods, horses, dogs, and falconers required much time, patience, and experience.

I had one surprising experience with a pair of gyrfalcons. During my studies, I experimentally fed gyrfalcons during the courtship period to see if I could encourage nesting in a year when ptarmigan were scarce. Every morning I would climb onto one of their feeding perches and drop off a dead ptarmigan. The pair quickly got used to me, allowing me to venture close to them at a time when they are typically sensitive to disturbances. No sooner had I left the feeding perch than the male would grab the food and court with it. Later when there were young in the nest, I rappelled into the site to band the young. Typically, this invasion into their nesting sites would incite an aggressive response (recall my first attempt to band youngsters), and even more hostile if the gyrs have had previous experiences with people. I expected the worst from this pair that knew me so well, but astonishingly, I was left alone, without so much as a fly-by.

Well camouflaged youngsters on their nest.

CHOOSING A NAME FOR THE GYRFALCON

The labels we have attached to the "gyrfalcon" have many interpretations (see Dalby 1965), but generally reflect their attributes, appearance, or disposition. Some insist that the word 'gyrfalcon' has its roots in Scandinavia from a Norse word, *geirfalki*, derived from *geirr*, meaning spear, and *falki*, meaning falcon. Perhaps this is a reference to the aerial strike from above, or perhaps the spear-shaped flecked feathers. Others insist the name stems from the old German *geier*, meaning vulture, a testimonial to the gyrfalcon's large size, from *giri* that means greedy, or from *gir*, meaning desire or eagerness. Still others stick to Latin origins of *gyrus* or *gyrare*, which means to rotate or to circle or to follow a curved path (notes from

Topsell 1658, cited in Shrubb 2013). Perhaps the name gyrfalcon is derived from the Greek word *hiero,* which means lord or sacred (Wood and Fyfe 1943; Cade 1968). In a 1555 publication, *L'Histoire de la Nature des Oyseaux*, Pierre Belon presumed that *gerfaulk* came from *gypsfalcus*, literally the grab-falcon. A later publication by Gesner in 1551 suggested that *girofalco* was synonymous with *Herodius*—the falcon trained for herons.

The Latin name for gyrfalcon has likewise been debated (Potapov and Sale 2005). Taxonomists have proposed a number of names for the gyrfalcon species within the Falco genus: rusticolus, canadensis, obsoletus, arcticus, islandis, sacer, canadicans, islandicus, and lanarius. Based on a lengthy review, Eugene Potapov and Richard Sale (2005) are of the opinion that Falco gyrfalco should stand as the valid Latin name for the gyrfalcon. As for subspecies, five to seven have been recognized based largely on predominant color across regions (Johnson and Burnham 2012): rusticolus (grey), intermedius (pale grey with some white), grbnitzkii (50% white), islandicus (grey, from Iceland), obsoletus (dark grey), and candicans (white) (Vaughan 1992).

There are many local common names for gyrfalcon: Labrador Falcon, Iceland Falcon, Greenland Falcon, Falcon Hawk, Snow Falcon, Winter Falcon, White Hawk, and of course Ptarmigan Hawk. In Russia, the gyrfalcon is referred to as *Krechet*, derived from a verb that describes the action of striking a forceful glancing blow. *Krechet*, says Georgiy Dementiev (1960), is believed to be of oriental origin, perhaps introduced into the Russian language by nomadic tribes from the east who at one time occupied much of Russia and to whom gyrfalcons were highly esteemed. The name shows up in Russian literature as early as the twelfth century. A slight variation, *Kereczet*, comes from the old Hungarian language, while the Polish word for gyrfalcon is *Bialazor*, "the whitish one." The Danes and the Norwegians call the gyrfalcon *Jagtfalk/Jaktfalk,* meaning the hunting falcon (Aegisson 2015). In the Middle East, the gyrfalcon is often referred to as the *Shunqar* (Phillott 1908), and in the Altai region as the *Turul*. In Greenland, the gyr is the *Kissaviarsuk*, and in Iceland, *Valr*. In the central east Yukon, the Kaska Dena have two names for the gyrfalcon: *Kusdli*, for the white one and *Togak'on* for the grey one (Charlie Dick, pers. comm). The Gwich'in Dena in the northern Yukon called the gyrfalcon *Kuw tsi chi* (Irving 1958). The Inuit also had a number of names for the gyrfalcon. Those in the eastern Arctic refer to the gyr as *Kigavic*, meaning the "the grasper" (Hohn 1969). In the western Arctic, the Inupiat have three names for gyrfalcons depending on its age, possibly because the feathers were of great value (Irving 1953). Youngsters were called *Atkuaruak*, meaning "like caribou mittens," juveniles were named *Kitgavikroak*, and adult gyrfalcons were referred to as *Okiotak, "the one who stays all winter."*

CHAPTER TWO

WHERE DO GYRFALCONS LIVE?

I DEVELOPED MY PASSION FOR GYRFALCONS IN the Ogilvie Mountains in the central Yukon. These northern subarctic mountains are some of the last true wilderness areas left in the world. Over most of these vast landscapes, the only obvious indication of human habitation is the odd camp, campfire, or woodcut. Here there are few roads, mostly game trails. The land appears to stretch forever under skies that are relatively pure, where the horizons are sharp, and mountain ranges are replicated as far as the eye can see, contrasting one another in a series of waves not yet violated by pollution. Because of the latitude, the mountains support vast areas of alpine meadows, below which fall expansive areas of shrub, unlike southern mountains where the boreal forest abruptly gives way to rock. During the abbreviated summer, the mountains are a mosaic of greens, etched with willows that mark the watercourses, and interrupted here and there with topographic breaks that produce cliffs and outcrops that are typically splashed with brilliant orange or yellow lichens. Closer to the ground there is the flush of wildflowers most often in shades of blue, white, yellow, and pink. Because the summer is short, they bloom all at once. The flowers are not typically robust and showy, but small, delicate, ornate, and prolific in this treeless environment.

The extensiveness of alpine and subalpine landscapes is pronounced in the fall when the green of summer is replaced by a kaleidoscope of colors: the tangerines of the dwarf birch leaves, the yellow of the willows, the burgundy of the blueberries, and the scarlet of the bearberry, with occasionally splashes of persistent green of the crowberry. This tapestry of colors is a prelude to the winter ahead.

Winter comes abruptly to the north. It comes in the way of snow, wind, a severe drop in temperature, and quickly diminishing daylight. There is something enchanting about the winter. The sun sits low in the horizon. The blues are deeper and richer, the images are crisper, and the twilight periods are drawn out adding further ambience to the land. Snow adds a finishing touch by softening the ridgelines and providing contrast to the sky. The purity and beauty of the subarctic landscape in the winter is a blessing that makes you forget the severe climate and those long periods of darkness.

Winter in the Ogilvie Mountains

Fall on the Mackenzie Mountain Barrens, NWT

Just as it comes suddenly, winter leaves in a miraculous way. The ice starts to form patterns of fractures and crevices, eventually transforming to water, and the patches of persistent snow are accentuated by melt-water glistening at the edges. All of this is brought to life with the sounds of spring: water trickling, the songs of birds announcing their intentions, and the chatter of small mammals including the spirited song of the whistling vole, released from the grip of winter. These events all happen before the drone of mosquitoes—the curse of this beautiful land.

Most of these subarctic regions where gyrfalcons principally reside are also the permanent homes of caribou and wolves, and on the craggy ridges, dall sheep. On the mountain slopes and basins also live hoary marmots as well as arctic ground squirrels, with collared pikas showing up in those enduring lichen-covered rocky talus slopes. All three species of ptarmigan find their home in the Yukon subarctic. The willow ptarmigan lives predominantly along the waterways and poorly drained tundra, where there are willows that are large enough to conceal them, the rock ptarmigan on slopes and ridges characterized by the prevalence of dwarf birch and other prostrate shrubs, and the white-tailed ptarmigan found sporadically along the rocky ridge tops. Where there are ground squirrels, ptarmigan, and marmots, the land supports wolverine, fox, and golden eagles. On the lower slopes in the high willows, the occasional moose can be found. Grizzly bears are also here. They are seldom seen but command respect even by the mere presence of their tracks in the soil. The bird life, although not impressive by southern standards, is noteworthy in that most arctic and subarctic birds offer one a chance to observe them in the splendor of their courtship attire when they are much more than music in the trees as is so typical of birds in the south. The breeding birds account for about 75-80 species in a mix of habitats. During their brief stay, they bless the tundra with their songs and flashes of color.

Although at home in the subarctic mountains, the gyrfalcon is typically associated with arctic tundra—the vast open plain north of the treeline. Most bird identification books illustrate the gyrfalcon sitting on a rock outcrop peering across the arctic tundra, showcased as a bird dwarfed by the immensity of a treeless wilderness.

But how are gyrfalcons distributed across this boundless hinterland of subarctic and arctic tundra? What features in these stark northern environments are they attracted to, or allow them to persist, and how do they allocate their range? These are particularly pertinent questions nowadays. Increasingly, there is more and more development pushing its way into the north and into the gyrfalcon's domain. Wildlife managers and conservationists are being asked to determine what's at risk and what has to be done to reduce the impacts on those species we deem as important. Gyrfalcons, rare symbols of the wilderness, pose a challenge to environmental watchdogs. The fact that they occur in the north far from the network of roads and communities limits the opportunity to study these large falcons to help reveal their distribution and ecology.

I suspect the basic criterion to explain gyrfalcon distribution is simple: where there are enough ptarmigan and adequate nest cliffs in open subarctic and arctic environments, there will be gyrfalcons. Indeed, the crucial factors for persistence of gyrfalcon populations in Russia, according to Vladimir Morozov (2011), are high numbers of prey and available nest structures.

There are exceptions. In coastal areas where cliffs are prevalent, it appears that gyrs can nest amid seabird colonies, or surrounded by wetlands, with less reliance on ptarmigan. They also nest in the high Arctic, where the occurrence of arctic hares year-round lessens their dependency on ptarmigan. Along some of the northern river corridors in Canada (Obst 1994) and over large areas of taiga forest in Russia where cliffs are absent, gyrs are also found nesting in trees. This is particularly so on the west Siberian plain of the southern Yamal Peninsula. Here biologists have found as many as 80% of breeding gyrfalcons nesting in trees, mostly along river corridors in a mosaic of taiga forest and swamps (Mechnikova et al. 2011). However, across most of their range they occur where there are ptarmigan, cliffs, and open spaces.

What we are more certain of is that gyrfalcons generally occur unevenly on their breeding ranges, above the treeline north of the sixtieth parallel, extending north even into the high Arctic. In North America, we find gyrs in the mostly treeless regions of Alaska, the Yukon, Northwest Territories, Nunavut,

Gyrfalcon habitat in the Ogilvie Mountains, Yukon

as well as Greenland (a few also breed in British Columbia). In Eurasia, they reside in Iceland, Norway, Sweden, Finland, and Russia (Potapov and Sale 2005).

Is there more to habitat requirements than simply ptarmigan and cliffs across arctic and subarctic landscapes? What specific habitats do gyrfalcons seek? After watching gyrs for many years, I am of the opinion that they take advantage of a variety of terrain features—valleys, ridges, draws, rock protrusions, permafrost mounds—to provide them surveillance perches or the advantage of surprise in the hunt. There is some indirect empirical support for this idea. On the flat tundra in Sweden, male willow ptarmigan during courtship were found to display at any time of the day, whereas in the mountains, courtship displays were reserved for the low light periods of dawn and dusk (Potapov and Sale 2005). Experimental trials with trained gyrfalcons hunting ptarmigan on flat tundra, although inconclusive, suggested that gyrfalcons were disadvantaged on flat terrain (Potapov and Sale 2005).

Ogilvie Mountains in the spring

I suspect other factors also influence the distribution of gyrfalcons. Quite likely latitude and the available hours of daylight dictate where they occur, as well as the persistence of snow and the height of the vegetation. What about ptarmigan? What is the threshold for the density of ptarmigan, below which gyrs do not occur? Does this vary between rock and willow ptarmigan? How about the presence of arctic ground squirrels? Or does the density of eagles, or perhaps snowy owls, or peregrine falcons, limit the

Gyrfalcon nest ledge, with chicks, Seward Peninsula, Alaska

occurrence of gyrs? No doubt a number of features influence where gyrs occur, but only in those areas where their basic needs are met.

In some areas, islands seem to attract gyrfalcons: Iceland, the Seven Islands off the Murman coast, Noyvaya Zemlya, and the Commander Islands, to name a few (Potapov and Sale 2005). Marco Polo (1324) noted in 1295 when describing Russia, "On that sea that there are certain islands in which are produced numbers of gerfalcons and peregrine falcons." Islands offer a number of advantages over inland habitats. Many are circumscribed by cliffs and devoid of mammalian predators. Ptarmigan are at an advantage with no ground predators and there are ample nesting opportunities for gyrs. Maritime cliffs also host seabirds, affording gyrs substitute prey (puffins, murrelets, etc). Thus, on suitable arctic islands, gyrfalcons occur more frequently, as claimed by Marco Polo.

How gyrfalcons partition these areas to meet their basic needs is all about nesting territories. Gyrfalcons, as with most birds, have adopted a territorial system that gives a breeding pair the exclusive rights to an area around the nest during the nesting period. In the Yukon, it appears that breeding pairs occupy a defendable area within a larger hunting range. They have exclusive rights around the nest site, excluding both nest competitors and predators.

Gyrfalcon nest ledge, with chick and addled egg, Baffin Island, Nunavut

Gyrfalcons are the polar bears of the bird world, at the top of the food chain in a stark environment. In the northern Yukon, the highest nesting density of gyrfalcons was one pair every 170 km^2 (Mossop and Hayes 1994). Similar densities are reported for Alaska, and the greatest cluster of active gyrfalcons was one pair every 175 km^2 on the Seward Peninsula (Potapov and Sale 2005). To put these figures into perspective, at their highest reported occurrence in pockets of the most favorable habitat in North America, there would only be about thirty-three pairs nesting in an area the size of Prince Edward Island, one of Canada's provinces. Remember, over most of their range, gyrs need both nesting cliffs and ptarmigan, neither of which are evenly distributed.

Across their entire range, gyrfalcons are most common in Northeast Iceland, where nesting pairs have been reported to reach densities of 137 km^2 (Nielsen and Cade 1990). Perhaps the juxtaposition of suitable coastal and inland habitats, and the absence of golden eagles, fosters high densities of gyrfalcons in Iceland.

Mackenzie Mountain Barrens, NWT

We know much less about the absolute number of gyrfalcons (nesting pairs as well as non-breeding adults and sub-adults), or where they spend the winter. I have seen gyrfalcons in the state of Washington and in the lower Fraser Valley in B.C. in the winter and have had leg-band returns from Alberta and Montana of youngsters that ornithologist Dave Mossop and I banded in the central Yukon. They were all juveniles, most of them females. I suspect gyrfalcons across Canada and Alaska wander south if they cannot afford to remain in the north. If ptarmigan are scarce or young gyrs have not mastered the necessary hunting skills, they wander away from their natal areas. Because young birds and females are probably the least accomplished as ptarmigan hunters, and perhaps less desperate for a nesting territory, they are the ones that are more likely to be found in southern areas. It is of interest that in these southern wintering areas, far away from their breeding range, there is some evidence that gyrfalcons, at least adults, establish well-defined hunting areas (Sanchez 1993).

In Alaska and Sweden, studies have found that young gyrfalcons tend to disperse in no particular direction, but often end up in coastal areas (McIntyre et al. 1994; 2009; Nygard et al. 2011). Gyrfalcons are regular winter tenants of the estuaries in B.C.'s Fraser Valley, and are recurrent visitors to a large interior water reservoir in South Dakota (Sanchez 1993). Coastal areas, estuaries, and large inland reservoirs often host high densities of overwintering waterfowl, shorebirds, and seabirds, further reminding us of Marco Polo's claim that gyrfalcons were abundant on certain islands in Russia.

There have been a number of observations of gyrfalcons perched on the broken ice at the edge of open water in the Davis Strait between Baffin and Greenland in the winter (Mosbech and Johnson 1999), apparently hunting seabirds and waterfowl. These ice-free oases in an otherwise ice-bound environment may be preferred, perhaps essential, areas for these arctic gyrfalcons. Also, it has long been known that white gyrfalcons (which do not breed in Iceland) occur in Iceland in the winter. According to falcon trappers during the eighteenth century, they were especially common during cold winters when there was more drift ice from Greenland (Jacobsen 1848, cited in Aegisson 2015; Salomonsen 1950, cited in Aegisson 2015; Poroarson 1957, cited in Nielsen and Petursson 1995). A recent study has substantiated anecdotal records regarding gyrfalcons on drifting sea ice. Kurt Burnham and Ian Newton (2011) obtained records of radio-attached gyrfalcons spending up to forty days at a time on sea-ice and icebergs amid seabirds in the winter. One bird spent about a hundred days at sea between Greenland and Iceland.

CHAPTER THREE

THE SUPERLATIVE HUNTER

THE ODDS ARE STACKED AGAINST AN ARCTIC predator. The Arctic is not a smorgasbord, but a simple food web. Hunting opportunities are even further constrained in the winter. For the few animals that remain in the north and are active in the winter, many go about their business obscured by a layer of snow and secure from most predators. Then there is the problem of natural population cycles—the boom and bust phenomena of many northern residents, most notably snowshoe hares, ptarmigan, and voles—which is periodically abundant, then scarce. Ptarmigan, collared lemmings, and hares also change their color so that they blend in with their background. They are white in the winter and brown in the summer. Additionally, most animals feed during dawn and dusk when the lighting hampers the ability of most predators to detect them. The gyrfalcon's body design, engineered to kill large prey, further constrains its hunting options. Hunters must meet a number of challenges—finding prey, capturing and subduing it, handling and transporting the carcass, and avoiding pirates. To find prey, gyrfalcons use more than one search strategy. They will hunt by soaring, particularly along ridges where they can use updrafts, but more typically, they will still-hunt from a perch with ample view, or quarter the terrain with a low elevation continual flight. Their greatest advantage is surprise and fast sustained flight. A common practice in the hunt for rock ptarmigan in Alaska was for gyrfalcons to attain high altitude to locate potential prey, then drop at high speed to less than a meter above the ground for a surprise attack (White and

Young gyrfalcon with ptarmigan kill

Weeden 1966). Tom Cade (1982) concluded that gyrfalcons typically catch avian prey after a long chase, often forcing them into the air to keep them aloft in order to tire them out. Likewise, Georgiy Dementiev (1960) commented on the gyrfalcon's remarkable power, speed, and persistence over long distances: "A gyrfalcon when cast onto a kite would sometimes bring it down to earth only after a pursuit of six miles, while an Icelandic gyrfalcon pursued a raven for about ten English miles." It is no wonder that gyrs used for falconry were traditionally flown with trained dogs and horses, or with watchmen stationed over the hunting grounds (T. Cade, pers. comm.), to help find and retrieve birds taken down many miles away from the first strike. But the chase is only the beginning.

Capturing prey is particularly hazardous for the gyrfalcon. Typically, large prey is harder to kill, and some fight back. Gyrs offset these risks with two lethal instruments—large talons with a particularly deadly hind toe, and the handy tomial tooth—to surgically dispatch its prey. The advantage of speed to deliver the decisive first blow helps. Repeated strikes and long pursuits also help by battering and tiring out the victim.

Gyrfalcons need open space to hunt. Unlike golden eagles, gyrfalcons will rarely fly into cover to attack their prey. For prey like ptarmigan and water birds, the odds improve greatly if the gyr can get underneath them and force them away from the security of the willows or water. If the falcon can get underneath its avian prey, it not only has the advantage in directing the aerial chase, but also tiring the prey out by keeping it aloft.

Male gyrfalcon with long-tailed jaeger, Seward Peninsula, Alaska

On one occasion, I inadvertently flushed a flock of pintails off a pond. This action prompted a gyrfalcon that was perched nearby to give chase. She quickly positioned herself underneath the flock. The ducks were left with only one real option: to out-fly the gyr. Pintails are great flyers. The fact that they range further north than most other ducks attests to this ability. The ducks that I flushed and the gyr in pursuit flew almost vertically upward, well out of sight of my unaided vision. My ten-power binoculars brought the drama into focus and allowed me to follow the chase upward until they were specs in the sky. Then, one pintail was separated from the flock. Perhaps it was a younger bird that became fatigued and realized he was losing the race. This pintail crumpled its wings and plummeted downward, with the gyr peeling off in pursuit. The pintail had only one chance left: to dive into the lake ahead of the gyr. But it was not to be. At about five meters from the lake where the chase began, the gyrfalcon struck the pintail, killing it instantly.

Although a mid-air kill is not uncommon, the gyr also "beats" its prey to the ground with repeated shallow stoops, often replicated over an extensive distance (Dementiev 1960). Battered and fatigued, the victim is eventually taken to the ground where the gyr can dispatch it with a well-placed bite between the neck bones. Gyrfalcons used for sport in past times were known to attack repeatedly and were gauged by the intensity and frequency of these strikes. Dementiev (1960) claimed that an average gyrfalcon flown by falconers in Russia would bring down a kite after about fifteen stoops, some attacking as many as twenty-five times. He further reported that in a letter written by Czar Alexei Mikhailovich in 1657 that one Siberian gyrfalcon brought down a kite after seventy attacks.

Gyrs will sometimes use distraction to improve their odds of a successful capture. On a few occasions in the winter, I observed gyrfalcons hunting in pairs, a male and a female seemingly working together. Recall my first encounter with gyrfalcons in the Ogilvie Mountains. Typically, one would fly ahead, low to the ground, the other just back and above. The lead falcon would flush the unsuspecting ptarmigan, leaving it to the second falcon to strike or get underneath it. No doubt this dual approach improves the odds for success. With its eyes on the first gyr, the ptarmigan may not spot the second gyr in time, or its response to avoid the first gyr may inadvertently put it at risk to the second and lessen its options to escape.

Gyrfalcons also better their odds with the help of other animals. In the winter when ptarmigan are in snow roosts, some mammals will inadvertently—or perhaps deliberately, as with foxes—cause ptarmigan to flush from their roosts. Once airborne, ptarmigan are vulnerable to gyrfalcons. Many times I have had gyrfalcons follow me as I skied across the tundra and have watched them often in the company of foxes. It may then be more than curiosity that attracts gyrs to dogs and skiers. It is likely a hunting strategy.

One Dena elder told me that her ancestors many years ago followed gyrfalcons to find caribou; that the gyrs would guide the hunters to the location of the caribou. Her story suggests that gyrfalcons seek out caribou, much like they do foxes, to capitalize on the opportunity where ptarmigan are evicted from snow roosts. Often moving in sizable herds across a snow-scape inhabited by ptarmigan, caribou will occasionally flush out ptarmigan from their snow-roosts. Some researchers have found that ptarmigan seek food from within caribou feeding craters, further bringing the two species together (Pedersen et al. 2006).

These behaviors of working together or using mammals to flush prey implies a level of problem solving that is quite unexpected of most birds. Canadian bird biologist Louis Lefebvre (2005) devised a way to score different bird species based on how innovative they were in their hunting and feeding habits. Not surprisingly, falcons and corvids (crows, ravens, magpies, jays) scored the highest of all birds tested according to the Lefebvre IQ index.

Hunting success is more than opportunity or chance encounter. The prey signals its fitness through behavior. Many times, I have watched wolves walking through groups of caribou. Usually they pass through, appearing oblivious, displaying no primordial urge to kill. Sometimes the wolves stop, watch, and occasionally stalk. On rare occasions, I have witnessed a chase. How do wolves know when the prospects of success warrant the chase? Most likely a caribou is signaling its fitness to the wolf by way of subtle behavior—posture, position, gait, or other such clues of vulnerability.

Gyrfalcons in all likelihood also use behavioral clues to gauge the vulnerability of their prey. When the gyr put those pintails into the air and pursued them, perhaps she was looking for a "weaker" bird. One falconer, John Lejeune, suggested that gyrs will often force a flock of ptarmigan into the air and "work them," seeking one that is less fit or evasive. A researcher in Russia also suggested that gyrfalcons are more apt to choose willow ptarmigan that are wounded or otherwise handicapped (Kalyakin 1989, cited in Potapov and Sale 2005).

A successful kill is the outcome of a combination of circumstances and aptitude—the skill of the hunter, experience to gauge its prey, and being in the right place at the right time. For the gyrfalcon, this is the consequence of superb flying skills, experience, and the fortuitous circumstance where the gyr finds its prey in compromising situations. These opportunities avail themselves more with young, less experienced prey—those that have yet to appreciate fully the threat of a gyrfalcon.

But the game is not over with a clever strike. Once the kill is made, there is the menace of pirates. Transporting a large carcass to the nest site is particularly perilous. Gyrs are handicapped when flying with large prey for the obvious reason that excessive weight impairs their ability to stay aloft and limits their agility. In these situations, gyrs have little hope of sidestepping an attack from other large predators, notably golden eagles, snowy owls, and ravens.

Gyrfalcon carrying a rock ptarmigan

White And Black; Artwork by Vadim Gorbatov

Once during April, I watched a male gyrfalcon, while transporting a ptarmigan to his nest site, intercepted by two ravens. The gyr, albeit the better aviator, was packing a ptarmigan that likely weighed 500 grams or so—as much as 40–50% of his own body weight. He lumbered along pretty well on a straight course, which in itself speaks to his strength of flight, but was severely compromised when defending himself and his kill. The two ravens took advantage of this situation. They methodically harassed the gyrfalcon, repeatedly diving from above. The gyrfalcon had no other option than to drop the ptarmigan. At this point, one raven continued to harass the gyr and the other immediately dropped from the sky and pulled the carcass into some thick willows. The gyr remained on his course and conceded the loss.

I suspect that this is a common occurrence, one of the inconveniences of bringing large prey back to the nest. The ravens appeared to have perfected this means of plundering by working together and dividing the labor. Not surprisingly, ravens, as with falcons, scored very well on Lefebvre's IQ test. But it is not only ravens that plunder from gyrfalcons. Snowy owls have been observed pirating from gyrfalcons (Fred Bruemmer, pers. comm.) as have golden eagles (Hardaswick and Christopher 2011).

Gyrfalcons, like other birds of prey, have a handy way of condensing the food they ingest to absorb only the most nutritious parts and avoid passing sharp bones through their gut. It makes for an efficient digestion, minimizes the risks of internal injury, and avoids compromising flight. This is achieved through a specialized foregut or gizzard that uses grinding muscles and acidic enzymes to break down the food. Indeed, gyrs and other falcons have a highly acidic digestive juice that is capable of even dissolving bones, says Tom Cade. The indigestible materials are expelled, or "cast," as a dense pellet of fur, feathers, and epidermal scales and keratinized layers of the beak. For the gyr, this digestive process is a way to get the most out of its food safely without hampering flight or risking internal injury. For biologists, the examination of castings has been a convenient way to estimate what gyrfalcons are eating without being on hand to observe every item consumed. Today, direct observations using sophisticated motion-activated cameras provide a more accurate estimate of diet (Robinson 2016). Nevertheless, the use of prey remains and castings continues to provide a general assessment of diet.

I used this now outdated technique of analyzing prey remains to determine what gyrfalcons were eating in the central Yukon. I collected prey remains and castings from fourteen different gyrfalcon nest sites from May through July, over five years. I sent the samples to the Zoo-archaeological Identification Centre in the Mammal and Ornithology sections of the National Museum of Natural Sciences in Ottawa, where, by breaking the samples into fragments and comparing with their reference collections, experts were able to distinguish by species and determine the minimum number of individuals in each sample, based on size and age of the fragments. Of the roughly 3,000 prey fragments that were assessed, 70% were identified to species. Only 6% of the fragments could not be identified beyond their taxonomic order.

Surprisingly, forty-three different species of birds and mammals were consumed by gyrfalcons in the central Yukon, yet almost 80% of the diet consisted of only four species: rock and willow ptarmigan (42%), arctic ground squirrels (28%), and upland sandpipers (10%). Shorebirds, of which there were nine

Willow ptarmigan in summer (above) and winter (below)

discernable species, made up 17% of the diet by numbers. Waterfowl (geese, ducks, mergansers, and grebes) were the next most common group to show up in the gyrfalcon diet. Surprisingly, voles, of which there were three identified species, occurred in just over 3% of the diet. More birds were eaten (63%) than mammals (37%).

An impressive study (Robinson 2016) in Alaska using sophisticated motion-activated cameras at active gyrfalcon nests through the entire brood period over two years (ten nests under surveillance in 2014, and thirteen in 2015) yielded similar proportions of major prey types to that in the Yukon. Of forty different discernable species brought to the Alaskan nests, ptarmigan accounted for 55% of the total number of prey, ground squirrels represented 26%, and shorebirds made up 19% of the gyrfalcon diet.

Although gyrfalcons ate everything from geese to voles in the Yukon, the species that showed up most in the gyrfalcons summer diet were roughly "ptarmigan sized"—the ducks were relatively small, the shorebirds were large. Harlequin and lesser scaup ducks were more apt to be taken than pintails and mallards. Upland sandpipers, whimbrels, golden plovers, and wandering tattlers were more often taken than least or spotted sandpipers, semi-palmated plovers, etc. Choosing the "right" size prey makes good sense for a specialized predator like the gyrfalcon. Too big is too hard to handle and risky to kill and transport. Too small is energetically inefficient as it would take too many individuals to compensate for the cost of capture.

Not surprisingly, ptarmigan were the essential component of gyrfalcon diet. Even in the summer when the spectrum of prey choice expanded with the influx of migratory birds and the emergence of hibernating mammals, gyrs continued to rely on ptarmigan, outnumbering all other choices. In nearly all gyrfalcon diet studies, ptarmigan have consistently shown up as the number one choice (Potapov and Sale 2005). It is no coincidence that the range of gyrfalcons in North America almost perfectly overlaps the range of ptarmigan, if white-tailed ptarmigan are excluded. After all, gyrfalcons are known by many as the "ptarmigan hawk."

I was unable to differentiate the proportion of rock and willow ptarmigan in the gyrfalcon diet. Both species occur in the central Yukon. However, Swedish researchers (Nystrom et al. 2006) used DNA analysis to determine the ratio of rock to willow ptarmigan in gyrfalcon diets and compared this against the estimated densities of the two species of ptarmigan. They found that the proportion of rock ptarmigan in the diet was indeed related to the amount of rock ptarmigan habitat, yet was over-represented in the gyrfalcon diet, as compared to willow ptarmigan. This implied a preference for rock ptarmigan. Because of their choice of open habitats with far less cover, rock ptarmigan may have been more vulnerable than their cousins (at least during this period of study). However, other studies suggest that where the two species of ptarmigan occur together, gyrfalcons prefer willow ptarmigan (Potapov 2011a). I wonder whether the proportion of young ptarmigan in the population at any given time influences the gyrfalcon's apparent choice of one species over another.

The white-tailed ptarmigan did not factor into the gyrfalcon diet in the Yukon—nor elsewhere, I suspect. They occur in small groups on high mountaintops in rocky areas and are widely dispersed.

Additionally, they are highly elusive. According to Tom Cade, they have an ingenious way of escaping from gyrfalcons by flying uphill and into rock debris slopes. White-tailed ptarmigan are simply too uncommon, too far away from gyrfalcon nests and preferred hunting areas, extremely difficult to spot, and use an effective evasive strategy, and so do not figure prominently in gyrfalcon diet.

Shorebirds, perhaps more than any other group of birds, have perfected flight and predator avoidance. Their ability to fly is exemplified by their remarkable migration habits. Some travel at least 11,000 km from their wintering areas to their breeding territories. The golden plover is known to cover a distance of 4,800 km in a single flight, even more impressive than the acclaimed flight of the arctic tern (Johnson et al. 2011). Shorebirds can generally outfly and outmaneuver most birds of prey and have devised behavioral habits to avoid predators.

American golden plover

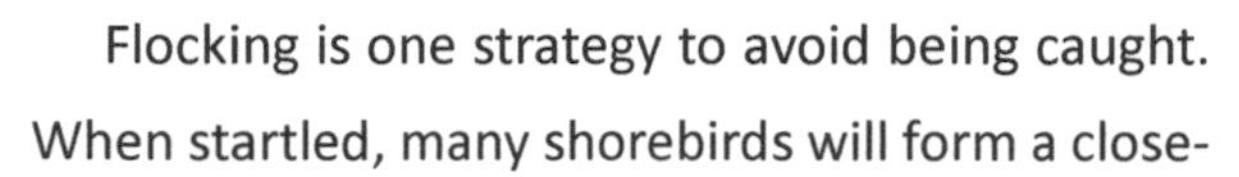

Flocking is one strategy to avoid being caught. When startled, many shorebirds will form a close-knit flock with remarkable cohesiveness, abruptly changing direction as a unit. This is much like a "school" of fish that respond to threats as a group. This strategy is thought to confuse the predator by impairing its ability to focus on one target.

Indeed, one study of Pacific Dunlin's along the coast of British Columbia found a strong correlation between peregrine falcon hunting success and Dunlin flocking behavior. As peregrine numbers increased along the coast from the 1970s through the 1990s, dunlins altered their behavior by substituting shoreline roosting with over-ocean flights during high tide when peregrine hunting was at a peak (Ydenberg et al. 2010). High tide flocking and sustained flights over the ocean as a mechanism to minimize the risk of peregrine predation, even at the cost of sacrificing fat reserves, was further evident by a drop in the kill rate by peregrines falcons when dunlins evacuated the shoreline for over-ocean flights (11% success over tide flats and ocean, as compared to 44% along the shoreline) (Dekker and Ydenberg 2004).

Shorebirds have other devices to avoid predation, as described by Peter Matthiessen (1994) in his book, *The Wind Birds*. The solitary sandpiper readily swims and can dive into water to avoid avian attacks. The spotted sandpiper is also good at this trick (Cade 1960). The Baird's sandpiper has a habit of "sneaking off" when disturbed by potential predators. The whimbrel, long-billed curlews, and black-bellied plovers are known to fight back against some of their avian predators. Some shorebirds also have a tendency to nest in clumps on the tundra. This is the same strategy that works so well for barren ground caribou; to give birth in large groups where there are more sentinels and where a cluster of newborns are exposed to relatively few predators, most of which are regulated by territories. Also, most shorebirds are cryptic in coloration to avoid detection.

Despite the shorebirds' many anti-predator adaptations, they made up 17% of the gyrfalcons' summer diet in the central Yukon, further affirming the gyrfalcon's remarkable hunting abilities. Likewise, Robinson (2016) found shorebirds to comprise 19% of the gyrfalcon diet on the Seward Peninsula, represented by nineteen identified species. Yukon gyrs preferred upland sandpipers and golden plovers, two species that are not confined to water and share the tundra with the gyrfalcon.

Wandering tattler

Of more than seventeen species of ducks nesting in the central Yukon, only a few types showed up in the gyrfalcon's diet. Gyrs appeared to prefer small diving ducks (those that dive for food) rather than dabbling ducks (those that bob for food). Diving ducks have their legs located further back than dabblers, making them better swimmers but less able to get airborne. Divers also have smaller wings in proportion to their body weight reducing the drag when they are diving. But this advantage in the water constrains their flight. It hinders their ability to get airborne and makes their flight more direct. Presumably, it is this decreased agility in the air that puts diving ducks at a disadvantage against gyrfalcons, if the gyr can get them off the water.

Long-tailed duck

Harlequin ducks (also known as "canyon ducks" here in the north) were the unexpected first choice for gyrfalcons in the central Yukon (based on the prey remains). They are not common. The eastern race is on Canada's list of species at risk, and the western race is not numerous, being confined to fast moving creeks in the north. It may be that Harlequins can be caught by gyrfalcons in the central Yukon more often than other waterfowl because they occupy habitats away from the lakes and potholes that afford most ducks security from avian predators.

Harlequin ducks

Arctic ground squirrels (often referred to as "gophers" here in the north) deserve special mention. More ground squirrel remains were found at gyrfalcon nest sites in the central Yukon, especially in July, than shorebirds and ducks combined. Bryce Robinson's (2016) research also underlines the value of arctic ground squirrels for gyrfalcons in Alaska. In one year of his study, 37% of the prey brought back to gyrfalcons' nests (51% by biomass) consisted of ground squirrels.

Arctic ground squirrels are the right size prey for gyrfalcons, small enough to take to the nest and large enough to make a reasonable meal. An elongate form that stores fat for hibernation makes them nutritious and perhaps easier to grab. Furthermore, ground squirrels live in colonies and produce many young—up to ten per female when times are good. In optimum habitat, in good years, they are abundant. They also spend considerable time in late summer outside their natal dens, feeding and collecting den material for the winter ahead and therefore vulnerable to gyrfalcons. I am convinced gyrfalcons are partial to ground squirrels. The energetic gains are good and they are comparatively easy to find and capture.

Predilection for ground squirrels was noted by falconer John Lejeune, who stated on his website, "I have found young gyrfalcons in a starving condition in Northern B.C. shortly after ground squirrels went to hibernate." Although it is common knowledge that young gyrs can survive on young ptarmigan,

Arctic ground squirrel

Lejeune's observations do imply a bias toward arctic ground squirrels, when young gyrs have a choice.

I suspect that ground squirrels in the Yukon, in most years, are a key resource to young gyrfalcons during the critical window between dependency on their parents and the onset of winter. Perhaps ground squirrels play an important role in the recruitment of gyrfalcons into the breeding population in the Yukon.

There were some surprises in the gyrfalcons' diet in the central Yukon. Two species—sora rail and ruffed grouse—were found in gyrfalcon prey remains but not known to occur in the area, or were extremely rare. These are what ecologists refer to as peripheral species: those that are found at the edge of their range, and likely outside, or peripheral to the areas to which they are adapted. Another rarity in the central Yukon that falls prey to gyrfalcons is the surfbird. These small upland shorebirds are extremely localized on tundra habitats high on mountain ridges (Frisch 1982) and likely infrequently encountered by gyrfalcons. I suspect the few that showed up in the gyrfalcon diet were killed when gyrs surprised them while ridge hopping in search of rock ptarmigan.

Voles are little mouse-like rodents that are seemingly unworthy of pursuit. Yet the frequency of voles in the gyrs diet in the central Yukon averaged 3%. Similar observations have been reported in northern Alaska (Cade 1960). Along the Colville River, five species of voles and lemmings were found in the gyrfalcons' diet, some years in significant numbers.

Arctic and subarctic voles and lemmings typically cycle in abundance and at times are superabundant. At such times, I have watched wolves (which typically kill things 8-10x their weight) hunting voles much as foxes do by stalking and pouncing, just as Farley Mowat (1963) described in his book, *Never Cry Wolf*. No doubt gyrfalcons, accustomed to hunting things much bigger, can also make sense out of hunting voles when they are periodically abundant. Indeed, in northeast Greenland, gyrfalcon populations were thought to track the four-year cycle of northern colored lemmings (Salomonsen 1950, cited in Potapov and Sale 2005). The shifts in the gyrfalcons diet during the breeding season appear optimal. Ptarmigan are the only prey available at the beginning of the gyrs' nesting season in most parts of the breeding range. Ptarmigan courtship, with its exuberant displays, flashy markings, and pre-occupation with mating, comes at an opportune time for gyrfalcons at the onset of the nesting period. As ptarmigan begin their own broods and the males again become cautious and cryptic, migrant waterfowl and shorebirds arrive. They are transient and there are many of them, and perhaps they are not wise to the habits and haunts of the gyrfalcon.

Red-backed vole

Arctic hare – a common prey of gyrfalcons in the high Arctic

Those that stay, separate from flocks and begin their courtship, which also comes at the cost of wariness. This situation is timely for gyrfalcons, providing a convenient substitution for the now cautious and cryptic ptarmigan. Soon ground squirrels become active and vulnerable to gyrfalcons. But this situation does not last. In late fall ground squirrels focus on the requirements of hibernation and are less venturesome, prompting gyrfalcons to shift their hunting effort toward vulnerable young willow ptarmigan, which are beginning to fly and as yet naïve. Not only are they vulnerable, ptarmigan are also easier to spot at this time of the year because they are molting and gathering into large groups. By shifting hunting effort away from ptarmigan during the summer, toward prey that is numerous and perhaps less experienced and easier to find and catch, ptarmigan are granted some relief from gyrfalcon predation. This reprieve means there will likely be more of them in the winter when the gyr has no other choices.

A very rare white arctic ground squirrel, perhaps a genetically distinct lineage that survived the last ice age in glacial refugium. The color may have been an adaption to avoid gyrfalcons in a landscape of predominantly snow.

CHAPTER FOUR

MORE ABOUT PTARMIGAN

IN THE YUKON, MALE WILLOW PTARMIGAN MOVE onto their breeding areas in March. They move away from the high willows that characterize their wintering areas to habitat where the willows are shorter and more widely spaced (Gruys 1993). Here they are more easily seen. Soon the males begin changing their plumage and advertising, eventually spacing themselves into territories. Their plumage and their behavior give them away to mates and predators both. With their contrasting chocolate napes and bright red eye patches, the male willow ptarmigan sit on exposed rock outcrops, atop mounds, or at the top of large willow trees, and chatter. They will also fly into the air, gaining a height of 7-10 meters while cackling, and then glide back to the ground. They are advertising to discourage other males and attract females. They do so with little apparent regard for the presence of gyrfalcons. The dull colored hens meanwhile remain rather inconspicuous. It's a clever strategy, says ptarmigan expert Dave Mossop. With the male ptarmigan drawing the attention of potential predators, females can concentrate on nesting without being noticed. There is some empirical evidence to support this interpretation. In Iceland, Arthor Gardarsson (1971) found that 33% of the male rock ptarmigan population he studied were killed in the spring at a time when few females died, and in his comprehensive study of gyrfalcons in Alaska, Tom Cade (1960) also found that gyrs selected male ptarmigan preferentially early in the nesting period.

The timing of ptarmigan courtship, with its exuberant, flashy males, could not be better for gyrfalcons. It is at this time when the male gyr is hunting for both himself and the female, and when ptarmigan are the only prey available to him. He must provide sufficient food to his mate so she will not vacate the site to hunt for herself. The arrival of shorebirds, waterfowl, and other migrants, and the emergence of ground squirrels from hibernation, is at least a month away. So, at a time when gyrfalcon nesting pairs are faithful to their nesting site, and rely on one hunter, they are favored by prey that draws attention to itself and is distracted. But is this fortuitous opportunity early in the nesting period simply a favorable circumstance, or is it a necessary condition of gyrfalcon reproductive success?

Male rock ptarmigan during courtship period;
note the head of the cryptic female in the lower right

Male willow ptarmigan during courtship period

One year in the Mackenzie Mountains in the Northwest Territories in Canada, I observed that the snow remained much longer than usual. Snow cover does not bode well for the mottled brown-colored female ptarmigan who rely on ground cover and camouflage plumage to avoid being spotted. Consequently, this lingering snow cover coincided with a two- to three-week delay in ptarmigan nesting (based on the age of the chicks I observed). Interestingly, young gyrfalcons failed to show up in the last half of August in a year when ptarmigan were abundant (presumably affording good prospects for gyrfalcon nesting success) and when gyrfalcons (at least in ptarmigan-rich years) can be seen frequently. Was this apparent unexpected reproductive failure of gyrfalcons somehow tied to the delay in ptarmigan nesting? By postponing their courtship and remaining vigilant and in larger groups due to unusual snow cover during the early part of the gyrfalcons' nesting period (a well-known fact according to Dave Mossop, and reported by Hannon et al. 1988), ptarmigan are less susceptible to gyrfalcon predation. Consequently, the ability of the male gyrfalcon to catch enough food for his mate and himself, at this important time, may have been compromised. Postponing courtship is not an option open to the gyrfalcons. To delay would mean fledging young at a time when their prospects for survival would be low. That season in the Mackenzie Mountains, when ptarmigan presumably remained cagey and gyrfalcon's hunting success was poor, many gyrs likely abstained from breeding. Consequently, few youngsters graced the barrens in the last half of August.

Female willow ptarmigan incubating her eggs. Note her camouflage plumage

Female willow ptarmigan with her brood

With the male drawing all the attention, the willow ptarmigan hen will molt in tune with the disappearance of snow, and eventually lay a clutch of 6-9 eggs. The eggs hatch to become precocial young, to face the world and the gyrfalcon. Once they have hatched, the cock will join his mate and the youngsters,

Ptarmigan pairs, rock ptarmigan (above), willow ptarmigan (lower left), white-tailed ptarmigan (lower right)

and they will move to denser cover where the willows are higher and more protective (Gruys 1993).

Parental dedication of the male varies among the species. In the willow ptarmigan, the male is a significant part of the family, keeping vigilant and defending the brood. On numerous occasions, I have had males fly at me when I happened to be too close to a brood and on more than one occasion, I have watched both male and female ptarmigan repeatedly fly directly at a northern harrier that presented a risk to their young. With the rock ptarmigan, the male is less directly attentive to the young. He stays in the wings, drawing attention away from the brood when danger threatens, while the female does the bulk of the parenting. The male white-tailed ptarmigan is nowhere to be found. Perhaps his presence alone represents far too much of a risk to the brood.

Willow ptarmigan in the winter

As the young ptarmigan grow and become more capable, we find a coalescing of families. Ptarmigan begin to form flocks and by early October, they move to higher elevations where they gather in large numbers of one hundred or more. Perhaps these larger groups give them the added advantage of more eyes, particularly important during their molt when they are less cryptic and their flight is compromised. At higher elevations, snow is expected earlier and so offers some camouflage when ptarmigan are molting. As snow becomes uniform across the land, ptarmigan descend, typically to river or creek bottoms where the willows exceed a meter in height (Gruys 1993). In some areas, ptarmigan remain in flocks throughout the winter, often faithful to specific areas, in numbers occasionally exceeding two hundred.

Snow is an important ally to ptarmigan by providing them with snow roosts to save energy and avoid being seen. They fly into soft snow usually near willows and quickly disappear from view with wind-blown snow covering their entrance. Here they remain in their little cavities, enjoying warmer temperatures, avoiding heat loss, and concealed from gyrfalcons, often for extended periods. A well-known Russian ecologist, Alexandre Andreev (1990), claimed that willow ptarmigan will spend up to twenty-one hours a day buried in the snow, out of the sight of predators. Similarly, in Finland there are reports that ptarmigan will spend all night and up to 80% of the day in snow roosts (Wingfield and Ramenofsky 2011).

This ptarmigan strategy of snow roosting works well against the gyrfalcon, but less so against foxes. With their uncanny sense of smell, foxes will detect ptarmigan in these snow roosts and gingerly approach them (Bergerud 1988). The challenge for the fox is to target and pounce on one before alerting others in the flock. With so many ptarmigan so closely spaced, the fox will often fail because the target is beyond other birds in the flock that are in his path of attack, but apparently undetected (Bergerud and Gratson 1988). Failure or not, the fox opens opportunities for gyrfalcons overhead.

Like a number of other northern creatures—snowshoe hares, lemmings, and voles—ptarmigan population numbers are cyclic. Some years they are abundant, then they become scarce, then abundant again. This pattern repeats itself over a relatively fixed period, or cycle, of about 8-10 years in the Yukon.

The causes of the ptarmigan to cycle remain unclear. Many ecologists have looked to food as the driving determinant for ptarmigan population cycles (Watson et al. 2000; Watson et al. 1998; White 2011). Changes in the amount of food or how it tastes or its toxicity, which in some cases changes with grazing pressure, constrain ptarmigan reproductive performance. Fewer young are hatched and still fewer survive when food is less available, less palatable, or toxic. Consequently, ptarmigan populations will decline.

There is the classical theory that increasing population size draws in more predators. The number of predators grows because they produce more young when prey is abundant, and they gravitate to areas where prey is plenty. With an increasing number of predators, the prey populations decline. But this explanation is not entirely convincing. When ptarmigan are abundant, it is not obvious that predators, usually regulated through exclusive occupancy of territories, can cause a significant dent in the ptarmigan population.

Ptarmigan snow roost

Parasitic worms have also been suspected as an agent in the cyclic phenomenon of ptarmigan (Hudson et al. 2003).

Another speculation to explain population cycles was proposed by well-known ecologist Dennis Chitty (1967). He suggested that changes in the interaction of individuals with different behavior, perhaps with different genotypes, influence population growth and decline. One behavioral type has the ability to produce many young but lacks the necessary care giving, which helps their young survive; the other produces few young but guards them well. Each behavior type has an advantage under different environmental conditions. When there are few competitors and lots of space and food, those that produce many young can quickly saturate the available space. When territories are fully occupied and competition is stiff, those able to secure and defend large territories are more successful, but at the cost of producing fewer young. Following this through, when ptarmigan are scarce, those individuals producing many young prosper, eventually saturating the available space, at which time the caregivers take over with their aggressive nature and ability to defend larger territories. The relative occurrence of these different types, possibly different genotypes, with their associated social habits, influences density and thus results in population cycles. It's a compelling theory, but to my knowledge, unproven.

Another theory, posed by Charles Elton back in 1924, correlated population cycles with sunspot activity.

More recently, climatic oscillations have also been implicated in the cyclic nature of ptarmigan (Watson et al. 2000). Here in the west, a ten- to forty-year climatic cycle, referred to as the Pacific Decadal Oscillation, appears driven by changes in sea temperature in the North Pacific and a corresponding change in the pattern of atmospheric pressure and the prevailing jet stream. The effect it has in the Yukon is periods of warming with heavier than usual snow accumulation, followed by colder, drier periods. Deeper and persistent snow may have an adverse effect on ptarmigan by reducing available cover in the winter and postponing the timing of laying.

I suspect it is not a single factor, but rather a series of circumstances that causes ptarmigan populations to crash: food (abundance and availability, quality, taste, digestibility), predators, parasites, disease, the prevalence of different genotypes, and demographic shifts, perhaps all influenced by climatic patterns. But because life is complicated, researchers have difficulty teasing apart cause and effect. The reasons some populations' cycle continues to remain a puzzle, albeit with compelling suggestions. What we do know is that for whatever reason, and likely a combination of factors, ptarmigan populations in many areas crash, most often as a result of fewer breeding pairs, lower fecundity and higher juvenile mortality (Bergerud 1988), and in doing so create an uncertain food supply for gyrfalcons.

Female willow ptarmigan in cotton grass

CHAPTER FIVE

ENEMIES & ALLIES

ALTHOUGH AT THE TOP OF THE FOOD chain, gyrs face at least three potential hazards: predators, pirates, and parasites. In the Yukon, gyrs are most antagonistic toward ravens and golden eagles. Ravens are both predators and pirates (as well as scavengers) and are commonly found where there are gyrfalcons. They will mob gyrs to steal their prey and perhaps raid unguarded gyr nests. To minimize these threats, gyrs will drive ravens away from the immediate nesting area, though attacks are typically restrained and of short duration (although I have found raven remains in gyrfalcon eyries, also reported by other observers).

Golden eagles are given no such latitude. In the Yukon, they elicited the most vicious response from gyrfalcons around the nest site. Without exception, resident gyrs would fly out to meet the intruder, gain height, and pummel the eagle repeatedly. The eagles would accelerate away from the hostile airspace, defending themselves on each attack by tipping upside down and exposing their talons. At no time did I observe an eagle initiate an attack. It was as though the eagle unwittingly flew too close to the gyrfalcon site, triggering the assault.

Golden eagles represent a sizable risk to gyrfalcons, primarily as pirates, but also for killing gyrfalcons that are heedless, brazen, or vulnerable. Tom Cade claims that golden eagles are the greatest threat to gyrfalcons participating in falconry in the western U.S. When an eagle spots a gyr taking a grouse to the ground it will fly directly to it. Typically, the gyr will willingly abandon its catch, but on some occasions, the gyr will not spot the golden eagle in time and it will be killed. Aggressive gyrs, unwilling to give up their kill when attacked, usually become fatalities. Surprise attacks by golden eagles toward careless airborne gyrfalcons have also been observed by Tom Cade, and reported in detail by Victor Hardaswick and Kent Christopher (2011). The gyrfalcons' hatred toward golden eagles appears to be universal (Potapov and Sale 2005). Is this hostile behavior toward golden eagles learned or genetically programed?

Bob Collins observed that captive bred gyrfalcons appear to be more susceptible to direct attacks by golden eagles than are passage gyrfalcons (those caught and trained after they had fledged). Bob conjectured that captive bred birds either fail to identify the threat, or respond too late or inappropriately.

Golden Eagle

His observations suggest that young gyrfalcons raised in the nest witness the interactions between their parents and birds of prey and learn to remain vigilant, identify the threats, and respond appropriately—essential skills for survival.

However, Kent Carnie, the founding director of the Archives of Falconry of the Peregrine Fund, observed behavior that suggested gyrs may be genetically predisposed to hate eagles. Although his observations were not of a genetically pure gyrfalcon, I believe they merit reporting. Kent had the opportunity to fly a "hard-imprinted" gyrfalcon-peregrine cross—a bird that had no contact with other birds from ten days old to well into September of its hatch year. When she began her training, the hybrid falcon was "weathered" by tethering her in a chicken-wire enclosure throughout the day until she was brought out to hunt in the late afternoon. Here in her enclosure she had many opportunities to observe birds of prey passing by. She was seemingly oblivious to harriers, buteos, merlins, and prairie falcons, regardless of how near they were to her. However, if an eagle was spotted anywhere on the horizon, she responded with obvious displeasure, loudly "cacking" for the entire period the eagle was in view. Despite never having observed or experienced other birds of prey during her upbringing, she was apparently wary or hostile toward eagles and indifferent toward other raptors (Kent Carnie, pers. comm.). Perhaps a gyrfalcon's fear of golden eagles is genetically hardwired, but their response is learned.

Young golden eagle

I did not directly observe predation on gyrfalcon chicks in the Yukon. However, I did find that at some sites, nestlings disappeared. At one of the sites under photo-surveillance, a brood of three vanished within a three-minute interval, and of 125 eggs I tracked during the study from nests that were not deserted, thirty failed to fledge young, some or all of which may have been a result of nest predation.

Here in the Yukon, eagles are the most obvious threat to gyrfalcon nest success. They are the most common large avian predator, and they arouse the most pugnacious response from gyrfalcons. For every active gyrfalcon nest in my area of study, there were at least three active golden eagle nests, some as close as four km from an active gyr site.

Normally nest predation is only possible when the female is absent. When the young are about two weeks old, the female gyr begins to share the hunting duties with her mate, which takes her away from the nest site. I suspect that when she is at home, no avian predator would deliberately approach her nest, but when she is away hunting, the brood is left at the mercy of nest predators. More nesting golden eagles in a gyr's territory likely increases the exposure of gyrfalcon youngsters to nest predators, including golden eagles as well as ravens, gulls, and jaegers.

Logically, the risk of losing their brood depends on the number of potential nest predators near-by and the length of time that the young are left alone and unprotected. The absence from the nest depends on how far adult gyrfalcons must go to find food and how long it takes them to achieve hunting success and return to the nest, which is influenced by the number and location of prey and the prevalence of pirates. Fewer ptarmigan or a lopsided proportion of older ptarmigan will compromise gyrfalcons' hunting success and extend their search time. Additionally, gyrfalcons presumably refrain from hunting within an area immediately surrounding a golden eagle nest. Eagles too are defending their young. As well as direct attacks to evict gyrfalcons near their nests, there is a high risk that eagles will spot and steal prey taken by gyrfalcons. This risk is well known to falconers. Often gyrfalcons will refuse to fly or they will leave the falconer's fist to find a safe place in the sky when there is a golden eagle overhead (Hardaswick and Christopher 2011).

Gyrfalcons, then, do not have unfettered access to their range. Where nesting eagles are common, gyrs are forced to extend their search farther away from their nests leaving their unguarded young at greater risk from nest predators. The effects of more encounters with eagles and longer periods when gyrs are away from their nest likely lead to higher gyrfalcon brood losses. Indeed, I found that 62% of the annual proportion of young gyrfalcons in the nest that died before fledging was correlated to the number of active golden eagle nests—fewer young gyrfalcons survived when there were more nesting eagles. Admittedly, there were few gyrfalcon brood losses. Nonetheless, a general pattern was evident.

Competition for food from golden eagles may also be unfavorable to gyrfalcons. In the Yukon, the diet of golden eagles during the nesting period was found to mirror that of gyrfalcons, with ptarmigan and arctic ground squirrels making up 56% of the number of prey items brought back to golden eagle nests (Hayes 1977).

Two Norwegian researchers, Kenneth Johansen and Arve Ostlyngen (2011), have also proposed that an abundance of nesting golden eagles may affect gyrfalcons. Similar conclusions were drawn from an interesting study of gyrfalcons on Alaska's Yukon-Kuskokwim Delta. Here, Brian McCaffery and his team (2011) compared three subpopulations of gyrfalcons. One of these subpopulations had the lowest nest occupancy rate and the lowest production of young, despite the apparent abundance of suitable nesting sites, as well as an abundance of ptarmigan and arctic ground squirrels. They explained this paradox by the extremely high density of golden eagles, elevated by the abundance of prey, which likely resulted in high encounter rates with gyrfalcons. They surmised that golden eagles could have an adverse effect on gyrfalcons by appropriating nest sites, stealing prey, directly killing adult gyrs, raiding their nests, and competing for food.

Golden eagle chick

In the Yukon, although gyrfalcon nestling losses were relatively insignificant when compared to breeding failures or abandoned eggs, the timing of these losses may be significant. When ptarmigan are at low density, there are fewer nesting gyrfalcons. This frees up space for nesting golden eagles. Those gyrfalcons, already compromised because of fewer ptarmigan, encounter more golden eagles. Because ptarmigan cycle over an eight- to ten-year period and gyrfalcons become capable of securing nest sites at the age of four or so (half way through the cycle), the few gyrfalcons fledged and able to survive when ptarmigan are few will occupy available nest sites when times are good. These premium chicks—the ones that contribute most generously to the gyrfalcon population—are at greatest risk to golden eagles.

Golden eagles and ravens, although potentially adverse to gyrfalcons as pirates and predators, build stick nests that serve gyrfalcons very well. In the central Yukon, 83% of all active gyrfalcon nests were on previously constructed golden eagle and raven nests. Perhaps gyrfalcons require these nest providers to remain in some parts of their range. A Norwegian biologist (Hagen 1952) suggested that Norway gyrs may be dependent on ravens for nest sites, and a Russian ornithologist (Kishinsky 1988, cited in Potapov and Sale 2005) suspected that the northern range of gyrfalcons in Siberia may be limited by the distribution of ravens. At the very least, I suspect the optimum nesting density of gyrfalcons in the Yukon would not be possible were it not for the nest building efforts of these archenemies.

Interactions with mammals were not usually hostile. Foxes were a common visitor to gyrfalcon nesting cliffs, probably drawn by the smell of prey remains. They would scavenge what they could and leave unmolested. Despite many such visits, I observed a direct attack by a gyr toward a fox on only one occasion, when the fox was near the nest.

On one occasion, a young Dall's sheep ram approached a perched gyrfalcon and shook his head a few times, which caused the bird to fly away. He then bedded down. The evicted gyr found a nearby perch, but a half hour later returned to provoke the ram. She dived repeatedly over the sheep, pestering him until he eventually got up, shook his head, and conceded the perch to find another lookout.

Raven on its nest with young

On another occasion, I observed an arctic ground squirrel navigating its way up toward the nest from below while the female gyr was incubating eggs. I suspect, like the fox, the ground squirrel was attracted to prey remains (arctic ground squirrels will eat meat (Cade 1951)). The ground squirrel continued, eventually surfacing directly on the nest ledge a mere meter from the gyrfalcon, an encounter that neither of them appeared to have expected. The gyr jumped up, and the ground squirrel toppled down the cliff in a controlled fall and escaped.

Red fox in the Ogilvie Mountains

When I approached a gyrfalcon nest during the incubation period the female typically responded in an aggressive way only when there was eye contact or I was walking directly toward the nest. When I walked parallel to the nest, my eyes averted away from the nest, the female was generally unresponsive, perhaps satisfied I had not seen her. Rappelling into a nest was quite another matter.

Gyrfalcon relationships with some mammals are what ecologists call commensal interactions. This is a situation where one species benefits at no inconvenience to the other. For example, recall that ptarmigan frequently use snow roosts in the winter where they are safe from gyrfalcons tucked away in their little snow cavities. The gyr benefits from the movements of foxes, humans, and caribou, which deliberately or inadvertently flush ptarmigan out of their roosts without regard to the actions of the gyrfalcon.

Caribou in the Selwyn Mountains, NWT

CHAPTER SIX

INVESTING FOR THE FUTURE

GYRFALCON REPRODUCTION IS INFLUENCED BY TWO SIGNIFICANT environmental challenges: a reliance on a single cyclic prey species and long winters. Gyrs have met these challenges through their choice of nests, the timing of laying, and a series of nesting decisions based on behavioral cues.

Large body size, although offering many benefits, has one major downside in the north. It comes with a long nesting period. There is no way around this for a bird that produces helpless young. To produce young gyrfalcons requires big eggs and a long incubation period.

The typical nesting schedule for gyrfalcons works as follows. The breeding pair engages in courtship activities for at least four weeks, then lays three or four eggs at 2-3 day intervals, incubates the clutch after the second egg for thirty-five days, then spends the next forty-seven days feeding youngsters in the nest. After the young leave the nest, the adults continue to feed the young, albeit less frequently, for another five weeks. Adding up the days, nesting gyrfalcons invest well over 140 days (almost five months) to produce independent youngsters. How do they accomplish this feat when winter prevails in the north from October through April, and summer lasts at best three months?

I suspect the timing of the nesting cycle is adjusted so that the young are fledged at the most favorable time. For most predators, the optimum time to leave the nest is when their prey is most plentiful and susceptible. It is a time when both the hunter and the hunted are developing and honing their survival skills. These survival skills can take many weeks, particularly for highly specialized hunters like gyrfalcons. If the falcons are fledged too late, they are up against prey that has had the benefit of time to acquire skills of vigilance and predator avoidance. The problem is even more exaggerated in the north where most of the potential prey migrates soon after fledging. For the Yukon gyrfalcon, the "right" time to fledge is the first half of July. It is the period when young arctic ground squirrels and ptarmigan chicks are naïve about predators and have not perfected the skills to evade gyrfalcons.

If they are going to meet this mid-July target, the gyrfalcon should be courting by early March and laying eggs in early April. In March and early April, it is still very much winter. Temperatures are typically

-20°C or colder at night. The land remains covered in snow and there is little evidence of life. Remember, this is a time when ground squirrels are still in their dens, the return of migratory birds is a month away, and the camouflaged ptarmigan are not always accessible, spending considerable time in snow roosts. So how do gyrfalcons accomplish this formidable task of initiating a clutch of eggs in the winter when the food supply is limited and timing is critical?

SELECTING NEST SITES

Gyrfalcons, like owls, do not construct their own nests. They seek suitable platforms, typically on appropriate cliffs, to lay eggs and raise their brood without interference from ground predators. Biologists call this an eyrie, or literally, "the nest of a bird that nests in a lofty place." There are exceptions. In some areas where cliffs are not available, gyrs will nest in trees, but across most of their range, gyrfalcons live in treeless environments where they require cliffs on which to nest.

Gyrs will use either a rock ledge or a stick platform. As I indicated earlier, 83% of all gyrfalcon nests I studied in the central Yukon were on previously built nest platforms constructed by either ravens or golden eagles. Some of these were entirely stick nests, while others were ledges that were extended with sticks to enlarge the nesting surface. One gyr nest was at the top of a two-meter pile of sticks, the work of golden eagles over many years. Where golden eagles or ravens are less common, rough legged hawks' nests provide a suitable substitute, although less stable.

Aside from a level platform on a cliff, are there other factors that influence gyrfalcon nest choice? Are there specific features of the nest and the nesting cliff that are important to gyrfalcons? Are the cliffs located at a particular elevation or is there an optimum cliff height? How about where the nest sits on the cliff? Do gyrs favor nests that face one direction or another? Are they more apt to choose a ledge or in an abandoned eagle nest? Is the configuration of the nest important? What makes a nest site desirable to a gyrfalcon?

Nest choice is important at these northern latitudes, particularly so to gyrfalcons that are obligated to initiate a clutch while it is still winter. In one year of my study, at the end of March, the temperature fell to -50°C. In all years -20 was normal, and snow and winds were common while gyrs were incubating eggs. Given the severity of the weather when gyrs begin nesting, do they choose nest site attributes to compensate for these harsh conditions?

In the central Yukon, as a starting point to understanding gyrfalcon nest choice, I measured nine features at fourteen nest sites, and I tested these against the total number of young gyrfalcons produced over six years (Barichello 1983). I assumed that those features that were prevalent at the nests that produced the most youngsters were the most important.

Yukon gyrfalcon nests were typically located at an elevation coinciding with treeline (at about 1,100 m). Where the gyr sites were equidistant, implying ample choice in where to nest, there was only 244 m difference between the highest and lowest altitude of nest sites. I suspect this was the optimum elevation. Too far above where the majority of the food occurs means an energy expensive trip up to the nest site from below. Too low brings the gyr into the boreal forest and away from its preferred prey. Ideally, the nest should be close to the food supply by distance and altitude. Golden eagles and ravens no doubt penalize gyrs for nesting too far from their food.

The height of the cliff or the direction it faced appeared to make little difference in gyrfalcon reproductive success so long as the cliff was higher than 50 m. Low cliffs simply do not provide the necessary height to prevent ground predators from gaining access to the nest, particularly when snow-drifts further reduce the effective height of the cliff.

The direction the gyrfalcon nests faced appeared to have no influence on gyrfalcon production. Southerly exposed cliffs benefit from spring warming and become snow-free earlier, but this would be preceded by freeze-thaw conditions that would produce icy conditions on the nest ledge. The benefit of advanced snow melt is probably more important to golden eagles as they nest later than gyrfalcons. Accordingly, 73% of the stick nests used by gyrfalcons in the Ogilvie Mountains were oriented south. Snow on the nest, although an inconvenience, will not prevent gyrfalcons from nesting. Gyrfalcons will scrape away snow on the nest to create a snow-free depression in which to lay eggs.

The type and size of the platform also had little influence over gyrfalcon productivity. A small raven nest produced no fewer young than a large eagle nest. Although gyrs did occupy golden eagle nests more

Female gyrfalcon on nest ledge

often than raven nests, it was not obvious that this was a preference. Golden eagle nests were far more common at the appropriate elevation than raven nests. Ravens nest earlier than golden eagles, perhaps discouraging gyrfalcons from contesting these sites.

One feature of the nest site seemed particularly important to gyrfalcons: how well it was protected from wind, precipitation, and other hazards. Nests with a roof over them produced more young than those that did not. The same was true for side walls. The additive effect of these two features, an overhang and sidewalls, yielded an even stronger relationship between nest protection and the number of young produced. Indeed, in a simple linear regression, 75% of the variation in the average production of young gyrfalcons across nest sites could be explained by the additive effects, weighted equally, of a roof and sidewalls (the horizontal angle of exposure of the nest). Nests above a threshold index of nest protection (roughly midpoint), produced an average of 2.2 young per nest over a six-year period, while those below the threshold index produced an average of 1.2 young per nest. Also, those nests above the threshold received eggs earlier on average (13 April) than the less protected nests (21 April). I suspect that nests that are tucked away in little alcoves or chambers on the cliff wall provide wonderful maternity wards. They not only provide comfort by shielding the young from rain, snow, cold winds, and extreme heat, they also protect the young from falling rocks, and reduce encounters with avian predators.

Gyrfalcon nest site with overhang and sidewalls

At one set of nesting cliffs in the central Yukon, there were two available platforms: one on a small ledge tucked into the cliff with an overhanging roof, the other a large exposed south-facing golden eagle nest. The golden eagle nest was used by gyrfalcons in the first few years of my study. I suspect it offered more room and a more secure platform. Eggs from a small nest are at much greater risk from rolling off if a depression cannot easily be scraped from the surface. Imagine how difficult it would be for the female coming and going and maneuvering around the eggs on a small ledge. It would be even more of a problem if the ledge were somewhat sloped. I decided to see if I could influence the gyrs nest choice by expanding the nesting surface of the rock ledge and eliminating any slope. Dangling from a rappelling rope in my sit-harness, with a free hand wielding a geologist's hammer, I picked away at the rock, eventually expanding the nest surface by about 20%

and creating a little depression. The next year (and in subsequent years), the gyrs abandoned the eagle platform in favor of the "improved" ledge with the roof over it.

Usually within the nesting territory there is more than one cliff, and presumably a number of nesting choices. On one cliff alone, there can be more than one suitable nesting platform. One might assume that moving nests year to year or periodically is a good strategy by allowing for a period of nest sterilization. Parasites are often associated with nests. It's an attractive environment for these little organisms, as it is warm and there are lots of decaying prey remains scattered about. It would seem logical that repeated nesting creates a more favorable environment for bugs and therefore a higher risk that young gyrfalcons will become the hosts of unwanted parasites. Consequently, we might expect gyrfalcons to move periodically, and more so after a successful nesting season, but gyrfalcons in the central Yukon moved infrequently. Perhaps the severity of the winter climate is such that it provides annual sterilization, regardless of whether the nest is utilized or not.

Movements by Yukon gyrfalcons to an alternative nest occurred only four times in the six years of my study (8% of all nesting attempts), and always following a nest failure in the previous season. It would seem that switching nest sites was a response to avoid circumstances that caused the nest failure. This was evident in the northern Yukon, where gyrfalcons vacated their nest sites the year following disturbance from helicopter overflights (Platt and Tull 1977). Apparently, the risk associated with this disturbance was enough to compel the gyrs to find an alternative site.

I suspect for many populations of gyrfalcons, nest choices are limited. Cliffs are not generally scattered evenly across the landscape and they vary in size and characteristics, or they may occur a long way from preferred hunting areas. Even finding an adequate-sized platform may not be assured. Some cliffs do not lend themselves to overhangs and side walls; at these sites the only available choice may be a nest platform created by an eagle, raven, or rough-legged hawk to meet a different set of nesting criteria. I suspect in the end, gyrs make do with what is available, with obvious preference for a site that offers security and protection from inclement weather, falling rocks, and predators.

The growth of orange lichens (associated with nitrogen-rich sites) and the accumulation of white guano at approximately seventy-five active gyrfalcon sites that I have visited suggest that gyr sites are used repeatedly over many years. But just how faithful are gyrs to these seemingly preferred sites? A study on Greenland used radiocarbon dating of guano samples to determine long-term use of nest sites. Incredibly, the oldest guano sample was estimated to be 2,360 to 2,740 years old and three others were over 1,000 years old (Burnham et al. 2009).

One question continues to puzzle me. Do gyrs compete for preferred sites? If there are features that improve the gyrfalcon's reproductive chances, and these are limited across the environment, is there greater competition for these limited and preferred nesting sites? Will gyrs vacate less favorable sites in an effort to secure more preferred sites? And if so, are the successes of nests more an expression of gyrfalcon fitness rather than the attributes of the nest sites themselves?

DEFENDING NEST SITES – TERRITORIES

In the Central Yukon where I studied gyrfalcons, most of their nests were evenly distributed, and typically spaced 12-14 km apart. In all likelihood, large territories are a prerequisite for a large predator with the task of raising a family in a subarctic environment where the potential prey is widely dispersed. But defending a large hunting area is inefficient and probably impossible.

An area with about a 2.5 km radius is aggressively defended here in the Yukon. Nonresident gyrfalcons passing within this airspace are followed, chased, or fiercely attacked. The attack continues until the trespasser egresses beyond the 2.5 km radius. The vigor of the attack is generally gender specific; males expel males and females expel females. Similar observations are reported elsewhere. Olafur Nielsen and Tom Cade (1990) observed a female attacking an intruding female, and saw resident males aggressively pursuing intruding males while escorting females away from the nest. They also observed that when a pair of vagrant gyrs passed through, the resident male directly attacked the male while ignoring the female.

Young gyrfalcons in the central Yukon appeared to be exempt from attack. Gyrfalcons, with the conspicuous darker streaked plumage and bluish feet, of either gender, spotted within the territory were pursued, but not with the hostility as with an intruding adult. Rather, they were escorted out of the nesting area. Joseph Platt (1977) in the northern Yukon observed the same pattern. It is probable that young gyrfalcons are distinctively colored until their age of maturity. The color distinction, combined with their submissive behavior, likely grants juveniles immunity when they innocently find themselves within defended airspace. After all, they are presumably of little competitive threat to the adult resident gyrfalcons.

I suspect few visitors will mount a serious challenge against occupied gyrfalcon nest sites. The intruders I observed appeared to find themselves inadvertently within an occupied gyrfalcon territory and the outcomes were not fatal. However, at one site in October I found a dead male gyrfalcon carcass; in all likelihood the victim of a gyr attack as evidenced by the severed vertebrae (remember the tomial tooth). Perhaps this was the outcome of competition for a nest site.

There is a practical explanation for why gyrfalcons in the central Yukon occupy nest sites 12 km apart, yet defend an area with only a 2.5 km radius. Clearly it would be futile, or impossible, to defend a very large area. Surveillance and eviction, even if possible, would be too time consuming and exhausting. At the same time, it is unlikely that gyrfalcons here in the Yukon can raise a brood within the confines of a 2.5 km territory, as the number of available ptarmigan in such an area would be insufficient in an average year. But why a 12 km spacing between nests in the central Yukon? Let us suppose a pair of gyrfalcons were to nest between two existing pairs, each defending a 2.5 km radius around the nest. The area would be near saturated with strongly defended nesting territories. Whenever a hunter would extend his search beyond 2.5 km from the nest, he would soon encounter another territorial male relentlessly defending his territory. In the search for prey beyond the defended territory, fighting would be common and everyone would lose. A 12 km separation between nests may simply be the distance that minimizes the rate of encounter of competing gyrs while allowing them an adequate area to hunt.

One study in Central-West Greenland lends some credibility to the notion that inter-nest distances are adjusted based on the probability or rate of interactions between neighbors (Johnson and Burnham 2012). They found that the number of offspring produced was inversely correlated to the distance between their nearest neighbors—fewer young were produced in nests that were close to one another. Close neighbors also had later laying dates. The logical explanation is that gyrs with close neighbors experience a higher frequency of encounters, and this negatively influences reproductive success.

I suspect, then, that where territories are more or less uniformly spaced, there is a tightly defended nest territory where confrontations are minimal, surrounded by a shared hunting area where gyrs avoid, but do not attack one another.

Nest spacing, however, varies across the gyrfalcon's range. Along the Colville River in Alaska, Tom Cade (1960) found the spacing between nests to vary from 3–30 km or more, depending on the year, yet contact between neighbors was infrequent. Cade suspected that, where pairs were close to one another, their territories were oval or pie-shaped rather than circular. Perhaps the shape of the territory is a local response to the clustering of available nest sites and perches, or where the prey is highly concentrated, as one would expect along these tundra river corridors.

Female gyrfalcon

FIDELITY TO EACH OTHER AND THE NEST SITE

Most birds form pair bonds during the breeding season. Each partner has an important role to play in raising a brood, and success requires commitments from both. But typically, loyalty to each other does not extend beyond the nesting period or carry over from year to year. However, some birds, including geese and swans, form a permanent bond. These birds, not unlike gyrfalcons, are large and have long nesting periods. A permanent bond minimizes the need to re-establish a relationship annually, and may be helpful for large birds with long nesting seasons and little time to court.

It appears that gyrfalcons also form long-term pair bonds. In addition to shared nesting duties, the associations may facilitate cooperative hunting beyond the nesting season. I observed gyrfalcons together in the winter on a number of occasions, presumably hunting. Perhaps the benefits of hunting together in the winter cultivate and reinforce the bond that is so critical in the nesting period. It is noteworthy that captive gyrfalcons exhibit strong social behavior toward each other, decidedly more so than other falcons (Hardaswick and Christopher 2011).

Where I was able to identify distinguishing features of individual gyrs, I became convinced that gyrs were faithful to the nest site year after year. At one site, the resident male in subsequent years had a unique, pronounced malar stripe. At another site, the unusual habit of the female gyr of perching on an arched willow stem was observed in successive years. I presume she was unique in this habit as no other gyrfalcon I observed used branches, but rather perched on rock outcrops. In consecutive years at a third site, I observed a female with unique color markings of alternating black, white, and grey, not unlike a spotted gyrfalcon. I presume these were all the same individuals repeatedly occupying the same nest sites in multiple years.

Other studies have also offered indirect evidence of individual gyrfalcon site fidelity based on color markings (Cade et al. 1998, Koskimies 1998, cited in Potapov and Sale 2005). One study found banded adults returning to the same nests (Nielsen 1991). Most gyrfalcon experts believe that there is a high fidelity to nesting territories (Potapov and Sale 2005), but I suspect they are also loyal to one another. Although I am unaware of studies that confirm long-term loyalty of pairs, an article by Marcel Gahbauer in the March 2002 issue of the Canadian Peregrine Foundation's "Talon Tales" stated that gyrfalcon pairs formed life-time bonds. Aegisson (2015) also stated that gyrfalcons are thought to mate for life.

COURTSHIP

Courtship is the next stage in bringing young into the world, and its function has been the topic of much discussion. Perhaps it is a mechanism to establish a pair bond, to synchronize the reproductive schedule, to convey information about fitness, or to stimulate hormones. Whatever the purpose, it is an important step in gyrfalcon reproduction.

To understand gyrfalcon courtship in the central Yukon, I observed nesting pairs at six relatively accessible nest sites for about nine hours a day over the entire courtship period. At two sites, I had little plywood "hides." At the other sites, I would park myself about 300 meters away and with the aid of a large binocular spotting scope would watch and record events. It was a tedious and tiresome way to collect data. In addition to long stretches of boredom when gyrs were perched and inactive, it was cold at these lofty observation posts, often -25°C, combined with wind chill. Cold enough that my fingers, even in mitts, would not always cooperate the way they should when called upon to record events. But there was some compensation. Gyrfalcons live in magnificent places with commanding views over a vast wilderness landscape, and I was able to enjoy the splendor of my "observation eyries" amidst these majestic mountains.

I spent 860 hours watching gyrfalcons court and recorded twenty-nine different activities (Barichello 2011). Some of these were normal maintenance activities like defecation, preening, fluffing up feathers (rousing), and casting pellets. I concluded that eight activities were strictly courtship-related, played out by both the male and the female in the drama that preceded nesting. These activities included the duration of time spent at the nest site, on the nest itself, and perched side-by-side at the nesting cliff, as well as the frequencies of food deliveries, nest visits, copulations, aerial acrobatics, and face-to-face vocal interactions. The events were often linked, with one triggering another.

Gyrfalcon pair on the nest ledge

In the central Yukon, the first obvious evidence of courtship display at the nest site begins in February and reaches a crescendo in the last half of March, just prior to laying. For the most part, it is a time when the male is the sole provider. The female simply bides her time, taking cues from the male, banking that he will return with food. The male, although eating from his kill, brings the bulk of the carcass back to his prospective mate.

Courtship is initiated by the male gyrfalcon. It is likely, as with most birds, that he gets the urge to court with the increasing length of daylight (Campbell and Lack 1985). His focus is on the nest. His primary task, it appears, is to get the female interested in this potential maternity ward, and is successful when he evokes the appropriate response from the female. Failing to do so, he works harder to woo her. Initially the male gyrfalcon seems preoccupied with the nest, making frequent visits as if to "show" it to the female, presumably as a way to stimulate her to come into breeding condition. In the central Yukon, his visits to the nest peaked three weeks prior to laying, and then dropped off as the female began expressing increasing interest in the nest. In a simple linear regression, 89% of the male's daily visits to the nest (measured across five-day intervals) could be explained by the female's loyalty to the nest—the longer she stayed at the nest, the fewer nest visits he made. Perhaps her complacency or coyness is designed to motivate the male.

Typically, nest visits and the excitement around the nest—copulations, aerial acrobatics, ledge displays, agitated "chipping" to one another—were stimulated by food delivery to the site. Often when the male returned with food, he paraded with it in front of the female and above the nest site in an impressive display, ending with an aerial food exchange, talon to talon. As these food deliveries increased, the female showed increasing interest in the nest and her mate.

I would sit for nine hours at a time waiting, watching, and recording. Most of the time it was uneventful: a female would perch, ever so often preen, but with little show of excitement. When food was brought to the nest, this all changed, with the gyrs talking to one another and actively displaying the full repertoire of behaviors. It was all I could do to keep up with pen and notepad.

How faithful the male was to the nesting cliff also seemed to be an essential piece of the courtship puzzle. I suspect his presence is a signal to the female that he is dedicated to defending his territory. Through most of February in the central Yukon, the nesting cliff was visited periodically, but neither the male nor the female lingered long. This all changed in early March, and by late March, just prior to laying, the male spent more than 70% of his time at the nest cliff, and upward of four hours at a time with the female.

Over time, the female too became more dedicated to the nest cliff. Her devotion appeared to reflect her mate's loyalty to the nest site and how often he delivered food. When the male was absent, coinciding with fewer food deliveries, the female, in the five days following, spent less time at the nest site. The statistical evidence is compelling: in a linear regression, 95% of the variation in the time the female spent at the nest site was accounted for by the amount of time that the male spent at the site in the previous five-day period. I suspect her loyalty to the nest site is influenced by hunger and the male's persistence at the site.

The number of copulations peaked in the week leading up to laying. Soon after, the female became much less energetic. Biologists call this "pre-laying lethargy." It is likely that the job of developing and laying eggs is very demanding and saps a lot of her energy. It is also possible that the calcium that is required for egg production is partially leeched from her own skeleton, as is the case in chickens (Taylor 1970). Diminished activity may be a side effect of calcium depletion. Perhaps hormonal changes prevent her from expending too much energy on any activities besides egg production. Either way, I observed a noticeable change in her level of activity up to five days before laying, consistent with studies elsewhere (Clum and Cade 1994).

This was not the case with the male. During the week of laying, he seemed even more excited. The time he spent at the nesting cliff increased, and his acrobatic displays were more frequent than in any other time during the courtship phase. I suspect these male exhibitions during the week of laying were not intended to entice the female to adhere to the nesting program, but rather designed to advertise his fidelity to the nesting site. Unwanted visitors would no doubt disrupt the nesting cycle at its most critical point, when the female is developing eggs, were not the male there to assume his duties to advertise and discourage intruders.

In the end, I concluded that courtship was largely about male advertising and female coyness in order to bring out the best in the male, played out with food as the primary motivator. Not surprisingly, imprinted captive female gyrfalcons during the breeding season demand to be fed and "do not graciously relinquish their food" whereas males are more giving toward their handler, to the degree that he "seems to act like his mission in life is to provide food for his mate" (Hardaswick and Christopher 2011).

EGG-LAYING, INCUBATION, AND HATCHING

If the male has done his job, and the female is well fed, preoccupied with the nest, and relatively quiet and inactive during the critical egg production process, she will lay 2-4 eggs (typically three or four, and rarely five) at 2-3-day intervals (typically about sixty hours between eggs (Platt 1977)). The interval between eggs is necessary because of the amount of energy and calcium required. A complement of four eggs represents as much as 14% of the female's weight (Clum and Cade 1994). It is impossible to produce four eggs all at the same time. Even chickens, bred for egg production and with ample food and calcium supplements, are hard pressed to lay an egg a day.

Gyrfalcon eggs are similar in size to large domestic chicken eggs. They're a bright cinnamon color at the time of laying, but will eventually fade and at the time of hatching the eggs are more of a cream color with a hint of rust.

In the central Yukon, the date of laying ranged from 24 March to 6 May. The median date of laying was 12 April. Nests that began early had more eggs: 67% of nests initiated prior to 10 April had four eggs, as compared to only 27% from nests where laying started after 24 April.

To develop and hatch successfully eggs must be carefully managed and maintained at a fairly even temperature and humidity. To assist in this endeavor gyrs develop paired lateral brood patches (Cade 1982). The brood patch is a featherless area on the belly and is equipped with many blood vessels close to the skin surface, which enhances heat transfer to the eggs. But the eggs also need to be periodically rotated, not just to maintain an even heat, but also to prevent the embryonic membrane from sticking to the shell.

In the Yukon, the female gyrfalcons that I observed did about 80% of the incubating. The males would offer help, usually when they brought food to the female. This exchange allowed the females a chance to feed, defecate, and preen their feathers.

Indeed, the females I studied had all the maternal instincts. They were much more attentive with the eggs than were the males. The females would carefully straddle the eggs, gingerly sit down on them, and shuffle them periodically, presumably to align their brood patches over the eggs. Occasionally they would get up, stand beside their clutch, and turn the eggs carefully with their beaks. It seemed incongruous to watch this large predator, capable of knocking geese out of the sky, so tenderly caring for her eggs. The males on the other-hand would simply plunk down on the eggs with little obvious care and maintenance. It is noteworthy that the males brood patches are also much less developed than those on the female gyrfalcon (Cade 1982).

Caring for eggs in the winter is decidedly challenging. At -20°C, which is common during the gyrfalcon incubation period, I expected that the eggs would quickly lose their viability if unattended for even brief periods. But in fact, gyrfalcon eggs are resilient to cold. At one nest site, I observed two ravens harassing a female gyrfalcon while she incubated her eggs. They approached the nest cackling and circling, getting ever closer to the incubating female. They were likely the previous tenants proclaiming their territorial rights and perhaps attempting to drive the gyrfalcons away from their stolen nest site. The gyr was tolerant for a while, but finally jumped off the nest and chased them. The ravens wheeled and raced away. I last saw them as they dove into a snow bank with the gyr close behind and gaining. The gyr returned but remained off the nest for a while longer, perhaps signaling to the intruders that she would not tolerate their return. She was off the eggs for over twenty minutes and the temperature was -20°C. All four eggs remained viable and produced four youngsters.

Some birds are known to lay a second clutch when the first fails or when they are afforded enough time and resources in the season to justify the extra effort of a second brood. There are few documented records of gyrfalcons having a second clutch in the wild (Potapov and Sale 2005). With such a short season and a long nesting period, a second clutch is unusual. However, a pair of gyrfalcons I observed on Baffin Island likely did lay a second clutch. It was late in the season—much later than usual to begin nesting—yet here was a nest with four bright cinnamon-colored eggs, indicating that they were recently laid. As it turned out, a few weeks prior to my surveys, two men were caught leaving Iqaluit with gyrfalcon eggs in a suitcase converted to serve as an incubator. The nest I observed with cinnamon-colored eggs was likely occupied by the pair that had lost their original clutch to poachers and were trying again.

The eggs hatch at slightly different times, probably due to intervals between laying and the necessity to begin incubation soon after the first egg is laid (Poole and Bromley 1988). All of the young typically emerge from the egg within fifty hours of one another, down-covered and with their eyes open. Soon after, they sit up and are ready to feed. The parents' job is now to feed them and protect them from brutal weather and predators. On hatching, the young are very vulnerable. They cannot move around well or regulate their body temperature. They are simply thirty-five-gram balls of fluff on top of two large feet, waiting for a meal.

RAISING YOUNGSTERS

Young gyrfalcons grow quickly. At ten days old they have a second layer of down and are beginning to regulate their body temperature (Poole and Bromley 1988). They double their weight in the first five days and achieve their most rapid growth over the next twenty days (Clum and Cade 1994). Feathers start to show up at about twenty-five days old, and by five weeks, the birds are nearly fully feathered. They remain in the nest for around forty-seven days. To achieve this rate of growth and develop feathers, young gyrfalcons require a substantial amount of food—about 50–59 grams per day (Nielsen and Cade 1990).

Raising a brood has other challenges besides securing enough food. Temperature fluctuations are a common feature of the north. Many animals are equipped to deal with unusually cool weather, but less so to accommodate excessive heat. Many seek habitats that offer them thermal relief. Moose, for example, use lakes and other water bodies in the summer to stay cool, "wallowing in huge content," as poet Robert Service (1973) so aptly put it. Caribou spend much of their summers on snow-packs and along wind-swept ridges depriving themselves of food to stay cool and avoid mosquitoes, bot flies, and warble flies. Grizzly bears dig summer dens for relief from heat. Gyrfalcon chicks are likewise prone to overheating. They are confined to a cliff, often exposed to the sun, and unless they have alcoves available to them, they have few options to avoid extreme heat. The problem is particularly hazardous in the first ten days after hatching when the young have limited ability to regulate their internal temperature.

Gyrfalcon chicks in their downy phase

But it is not just heat. The chicks also face cold weather conditions. In the north, the summer weather can change dramatically, particularly at high elevations where gyrfalcons live. Twenty-degree swings in temperature are not unusual, and wind, which exacerbates the cold, is more extreme around cliffs. The nest cavity can moderate the extremes in temperature, but that is not always enough. In these situations, brooding becomes important.

To explore these challenges of caretaking for nest-bound young, I set up primitive little 8mm film cameras (before the age of digital cameras), programmed to expose one frame every three-minute interval, at six different gyrfalcon nests. I left these remote cameras to record over 40,000 frames of information over three nesting seasons—roughly eighty-three days of surveillance. This technique was not as simple as it may sound. It was always a challenge to find a ledge on which to place the camera: close enough to see into the nest but not too close to disturb the birds. Also, most nests were on sheer cliffs, making camera placement difficult. The sites where camera location was feasible were some distance off the road and entailed a long trek to get to the cameras to replace the film every three or four days. For some of these sites it was over a 16 km hike there and back. Then there were the problems of keeping cameras warm and dry so they would continue to function, and making them safe from animal damage. As a solution, I built little plywood boxes with Plexiglas windows to house the cameras, complete with battery and little thermostat-controlled toaster elements to keep things warm and functioning.

Female gyrfalcon brooding young during a blizzard

Brooding was a common activity from the time young hatched until they were about two weeks of age, roughly the period when they were balls of down. Tallying up the number of photographic frames where the female was observed brooding the young as a percentage of the number of frames exposed, and comparing this week to week, I found brooding was evident 80% of the time for the first two weeks, then declined sharply. At twenty-one days—about the time feathers are beginning to develop—there was virtually no brooding. The timing is perfect. The female ceases brooding when the young require more energy to produce feathers and become more boisterous and difficult to brood. This frees up the female to begin hunting to supplement male food deliveries.

How often are nest-bound young gyrfalcons fed, and does this change as they grow? In the central Yukon, adults fed or left food for their young on average every 7.5 hours, equivalent to 3.2 feedings per day. This timing was consistent for the first four weeks. Food delivery rate decreased slightly after the fourth week, consistent with findings in Alaska and the central Arctic (Bente 1981; Jenkins 1982; Pool and Bromley 1988). The obvious explanation is that the adults were delivering larger prey and therefore fewer deliveries were required. Arctic ground squirrels weigh about 700 grams as compared to ptarmigan that typically weigh about 500 grams (Robinson 2016), and more ground squirrels showed up in the diet as the summer progressed. But perhaps there is another explanation. My cameras may have failed to pick up small prey deliveries. Normally parents do not linger at the nest too long when the young get big and lively. If the eyasses are rambunctious, they can cause injury, and when they are hungry, the nest is mayhem. It is possible that my cameras failed to capture smaller prey items that were quickly pulled to the back of the nest and devoured.

Gyrfalcon feeding chicks, Seward Peninsula, Alaska

One feeding phenomenon intrigued me. In my study, the female did most of the feeding. She carefully broke apart small pieces of a carcass and delicately offered it to a youngster. She seemed unconcerned about which of her young she was feeding. She simply provided the tidbit to any gaping mouth in front of her. Although apportioning food evenly between youngsters has been reported in studies in Alaska and the north Yukon (Bente 1981; Platt 1977), I suspect this rationing of food is not deliberate but occurs because as one youngster devours a morsel or becomes satiated, he or she is less apt to beg for food, leaving an opportunity for the other chicks. Random feeding is selfish and sometimes fatal for weaklings, but a sensible arrangement when there is not enough food to go around. It is better that the strong chicks survive at the expense of the weaklings, rather than all chicks suffering and dying because of insufficient food. Dividing food equally would seem to be fair, but it is a bad evolutionary arrangement when food is not abundant because it would not favor the strongest individuals.

What was surprising to me, initially, was that the female seemed obsessed with the size of the tidbit she was feeding her chicks. If the piece of meat she had torn from the carcass was "too big," and one of the youngsters grabbed it, she would wrestle it away from the chick so she could tear it again into a smaller piece. Perhaps choking is a frequent hazard when the nestlings are small, or perhaps there is a greater chance that there are bones in these bigger pieces, which would pose a threat to hungry youngsters. Perhaps mothers have "learned" through natural selection to avoid feeding the young large pieces of meat. A friend of mine, Jannik Schou, who spent long hours watching and photographing gyrfalcons in the northern Yukon, shared an interesting observation with me. He observed a chick with a bone lodged in its throat and unable to feed. Returning to the nest three days later, he found the chick to be emaciated but the bone was gone. Despite its weakened state, the chick recovered and fledged with his siblings.

Gyrfalcon with chicks, Seward Peninsula, Alaska

Removing food remains from the nest platform may be another way to minimize the risk of young gyrs ingesting bones. The researchers in Greenland who put nests under camera surveillance reported that gyrfalcon parents removed about 21% of the prey remains they brought to the nest after the young had fed, of which 39% contained bones (Booms and Fuller 2003b). Presumably, this removes the temptation of young birds devouring bone fragments, as well as removing materials that could attract scavengers or increase the risk of diseases.

The nestlings occasionally fought over food portions, but the interactions between them were non-aggressive. I did however watch a youngster innocently be catapulted out of the nest to his death. Older youngsters will mob their mother when she brings in food, and on this occasion, a little too much excitement to get first dibs on the food resulted in one young eyass accidentally getting pushed off the nest. Perhaps he was a little younger, a little weaker, or less driven by food than his nest mates. I suspect he died when he hit the ground some forty meters below. The mother suspended feeding for a second, then resumed feeding the remaining young, apparently oblivious to what had happened. But surprisingly, once she had finished distributing the food, she flew from the nest directly to the bottom of the cliff to inspect the young. Had he been alive, I wonder whether she would have returned him to the nest.

A motion-sensitive camera in Alaska has indeed recorded a gyrfalcon moving its youngster (Robinson et al. 2017). An adult female gyrfalcon was observed with one of her youngsters in her beak at the edge of a collapsing nest. A later observation found the youngster alive in an alternate nest platform five meters from the original nest. It was obvious that the adult female had rescued her youngster from the collapsing nest and carried it unhurt to an alternate nest close by (Robinson et al. 2017).

Yukon gyrfalcons were not on a particularly rigid feeding schedule as is the case with many animals. The feeding rate was highest between about seven in the morning to about 10 o'clock in the evening, peaking at about four in the afternoon. There was a well-defined lull in feeding from about 11 pm to 4 am. In the north, it is light for twenty-four hours a day during the summer—the land of the midnight sun. Dawn and dusk merge into one another and are of brief duration. For gyrfalcons, perhaps one time of day is as good as the next to hunt for food. The sun is below the horizon in the central Yukon only from about midnight to three in the morning, a time when gyrs either hunted less or were less successful.

The timing of food deliveries in the Yukon was similar to that observed of gyrfalcons under surveillance in Greenland (Booms and Fuller 2003b). There, food drops similarly showed a subtle peak at about 4–6 pm, and likewise tailed off after 10 pm with a pronounced lull from midnight to 4 am. Ptarmigan have possibly adapted to gyrfalcon hunting schedules. In Sweden, rock ptarmigan were found to be most active feeding from 11 pm to 3 am (Nystrom et al. 2006), coinciding with a lull in gyrfalcon food deliveries.

Gyrfalcon chick near fledging.

FLEDGING AND GAINING INDEPENDENCE

For young gyrs, learning to survive and hunt requires experience and takes time. They must master fight and develop hunting skills, all the while remaining vigilant and responding appropriately to potential predators. Parents are a necessary part of this training, feeding the youngsters, protecting them, and possibly providing them a model to emulate.

Just prior to leaving the nest, the young spend considerable time flapping their wings at the edge of the nest. They are building flight muscles and "testing the air," waiting for the appropriate time for lift off. More often than not, they leave the nest by catching the wind and soaring a short distance rather than immediately being capable of controlled flight. They then spend a number of days hopping from perch to perch and intermittently soaring. Eventually they "get their wings" and can sustain powered flight. They are now capable of flying out to meet parents when they bring food back to the nesting cliff, giving the early fliers an advantage over their nest-bound younger siblings. Perhaps hunger coaxes the remaining young into the air, a practice used by falconers to encourage captive birds to hunt—but these young gyrs still cannot fend for themselves.

It takes at least five weeks on the wing for young gyrfalcons to be able to hunt well enough to feed themselves without needing assistance from their parents. On the Mackenzie Mountain Barrens, I would begin observing young gyrfalcons in the third week of August. This high elevation plateau is an ideal training ground for young gyrfalcons. There are many pockets of ground squirrels, good habitat for ptarmigan, and lots of open space far away from active gyrfalcon or eagle nests. The young are likely leaving the nest in the second week of July (assuming the timing is similar to that in the central Yukon) and showing up on the Barrens when they have ended their dependence on their parents, roughly six weeks later.

Gyrfalcon chick, near fledging, Ogilvie Mountains

My speculative conclusions about the timing of independence are supported by research in Alaska. A team of researchers (Carol McIntyre, David Douglas, and Layne Adams 1994) put satellite transmitters on young gyrfalcons within ten days of fledging and tracked them for a month. They reported that these young spent an average of 41.4 days within their natal areas after fledging, then dispersed between 20 August and 2 September. Other studies have largely confirmed these findings (Potapov and Sale 2005). The young need approximately four weeks to master flight, and another two weeks before they can hunt for themselves.

WAITING FOR A NESTING TERRITORY

One intriguing facet of the life history of large northern raptors is the delay between independence from their parents and the time they acquire their own nesting sites. Few other animals experience such a delay—independent, sexually mature, but foregoing or excluded from nesting for lack of a defendable territory of their own. Ecologists refer to these non-breeding individuals as “floaters”: sexually mature individuals that are surplus to the breeding population and “float” or wander around until they can acquire a breeding territory. The evidence for this surplus population of non-breeders comes from experimental removal of birds holding territories. Territorial birds that are removed are quickly replaced with new tenants.

The most common explanation for the existence of floaters is that there are only so many territories, and established territorial pairs prevent surplus adult birds from breeding. This situation implies that territory size or suitable nesting locations limit the number of breeding territories. Another explanation is that it takes time to develop the level of fitness required to occupy and defend a nesting territory and satisfy a mate. It may be that those that put off the business of nesting until they are more capable, produce more young over their lifetime than is afforded to those that begin early and inevitably fail, "burn out," or are eventually evicted. For long-lived birds like gyrfalcons, this tradeoff between current and future investment may favor those that choose to wait. Either explanation rests on the assumption that reproduction is a costly endeavor, achieved only with a suitable nesting territory, and a relatively high threshold of skills necessary for success. Consequently, some adults are excluded from breeding—the floaters.

Young great horned owl

One noteworthy study of the great horned owls in the Yukon by Christoph Rohner (1997) revealed the fate of surplus birds, and it perhaps best mimics what I would expect of gyrfalcons. Although nesting in the boreal forest and not in open areas, great horned owls, like gyrfalcons, prey predominantly on a single cyclic prey species (snowshoe hares), and they nest very early. By tracking thirty fledgling owls with radio transmitters, Christoph was able to draw a number of conclusions. He found that 30% remained near the area where they were hatched and occupied an area five times that of their parents. They ventured periodically into occupied territories, but spent much of their time in the peripheral areas (presumably to avoid fatal encounters with the territorial occupants). There were many of these floaters when snowshoe hares were abundant (almost one for every two territorial occupants), but they suffered high rates of mortality or left the area when the hares crashed, at a time when territorial occupants were unaffected. I am guessing that gyrs have an even more demanding existence. They rely on avian prey that are typically at lower densities than snowshoe hares, in what is perhaps an even less forgiving environment, at higher elevations and higher latitudes.

REPRODUCTIVE AGE AND LONGEVITY OF GYRFALCONS

There are no studies that I am aware of that have confirmed the age at which wild gyrfalcons are capable of raising a family or how long they live. Captive-bred gyrfalcons are known to begin breeding in their third year (Platt 1977; Palmer 1988; Cade et al. 1998), or even in their second year (Peter Devers, pers. comm.). I suspect wild gyrfalcons need more time to attain the nutritional state required to initiate breeding, as they are more active and their food supply is less certain, as compared to well-fed captive gyrfalcons.

The life span of gyrfalcons has also been conjectured but with little evidence. At one site monitored in Sweden for fifteen years, an individual male occupied the site for the whole period, suggesting a life span of at least eighteen years (Lind and Nordin 1995, cited in Potapov and Sale 2005). However, the bird was unmarked, raising uncertainty as to whether this was the same individual. The oldest ringed bird recaptured was a twelve-year-old male (Cade et al. 1998). Based on these meager observations, it can be assumed that wild gyrfalcons live to at least twelve years of age, and perhaps as old as eighteen. Captive bred gyrfalcons can live longer, I suspect due to their comfortable lifestyle. They are well fed, expend less energy, and are free from many of the demands and hazards that face wild gyrfalcons.

CHAPTER SEVEN

HOW DO GYRFALCONS COPE WITH CYCLIC PREY?

ALTHOUGH PTARMIGAN ARE GYRFALCONS' MAJOR FOOD SOURCE, the cyclic nature of their populations has created a considerable challenge. I was fortunate in my studies: Mother Nature provided me with the "experiment" to investigate this puzzle. For three years, ptarmigan were abundant, then they suddenly crashed and remained relatively scarce for the three subsequent years. This sudden collapse in the ptarmigan population represented a six-fold change. By measuring gyrfalcon reproductive success against this change (and other attributes of the environment), I was able to describe how gyrs respond to this sudden drop in their primary prey, at least for one cycle in the central Yukon (Barichello and Mossop 2011).

The effect of a decline in the number of ptarmigan on gyrfalcon reproductive success from fourteen nest sites under observation in the central Yukon was indeed striking. In the three abundant years, ninety-five young gyrfalcons left the nests, but during the ptarmigan recession, just thirty-five young were fledged. Using statistics to quantify this response further, I found that 92% of the year-to-year variation in the average number of gyrfalcons that were fledged could be accounted for by the abundance of ptarmigan.

Although convincing, I was curious about one apparent anomaly. During a year when ptarmigan numbers remained high, there was a drop, albeit slightly, in the number of gyrfalcons that left the nest. Perhaps this slight decline in gyrfalcon production was associated with the proportion of young ptarmigan in the population. Dave Mossop had randomly sampled ptarmigan within the gyrfalcon study area in late winter and found that the proportion of young ptarmigan had indeed dropped during the year that gyrfalcons faltered. In fact, I found that 87% of the variation in gyr production in the three years when ptarmigan were abundant could be explained by the proportion of young willow ptarmigan in the population.

Four gyrfalcon chicks on a nest built by ravens

The importance of young willow ptarmigan to gyrfalcon reproductive success has been validated in Sweden and Norway. In Sweden, Falkdalen and associates (2011) found that a high proportion of young willow ptarmigan in the winter had a sizable role in gyrfalcon breeding success in the upcoming season. Similar conclusions were drawn by Selas and Kalas (2007).

Another obvious feature of this relationship between gyrfalcon reproductive success and the supply of food was the timing of laying. When ptarmigan were abundant, most gyrfalcon pairs began a clutch prior to 12 April, yet few pairs met this deadline when ptarmigan were scarce. In a statistical analysis, 84% of the year-to-year variation in the median laying date (an equal number of nests were initiated before and after this date) could be attributed to the number of ptarmigan. Gyrfalcons had a head start when ptarmigan were abundant.

Gyrs that postponed laying in the central Yukon did not fare well. Most of the clutches that were started prior to 12 April yielded young (93%), but only 38% of later nests produced young, and because late-fledged young are up against a more experienced prey population, fewer were likely to have survived.

Not surprisingly, a significant decline in their major food supply, ptarmigan, had a drastic effect on the ability of gyrfalcons to produce viable young. But where in the reproductive cycle did the failures occur?

EARLY FAILURES

In the central Yukon, most gyrfalcon reproductive failures were evident early in the nesting cycle. The pairs that failed to produce young more often than not failed to breed or abandoned their eggs within a few days after the clutch was complete—91% of all nest failures were due to these factors. In years when ptarmigan were abundant, gyrfalcon eggs were produced in 89% of all active nests, yet in ptarmigan-poor years only 58% of occupied sites yielded eggs. Of those eggs laid in the ptarmigan-poor years, 33% were deserted soon after they were laid. If eggs were hatched, there was little year-to-year difference in the number of young produced. Gyrfalcon pairs, it would seem, cut their losses early so as not to invest heavily in a risky venture. Those that did succeed in incubating a clutch of eggs generally fledged a normal-sized brood.

Willow ptarmigan in the fall; all eyes on a gyrfalcon overhead

But there were other subtle adjustments linked to food supply. The number of eggs in the nest varied somewhat, presumably related to ptarmigan numbers. In the year when ptarmigan were most common, 62% of all clutches I observed had four eggs, while in other years four-egg clutches were found in fewer than 40% of all clutches.

The ability of the female to mobilize the necessary nutrients to produce eggs is a prerequisite to laying. For gyrfalcons, I suspect there is a high nutritional threshold to begin laying, but that threshold must be sustained during the laying period to produce the full complement of eggs. For some females in some years, it may have been possible to begin a clutch, but because they could not maintain their nutritional state due to fewer feedings, only three eggs were laid. More convincing was the fact that clutch size was linked to nest failures. Deserted clutches were smaller (an average of 2.9 eggs as compared to 3.3 from successful clutches) and were laid later (average laying date was 30 April as compared to 12 April from those not deserted).

Deserting the clutch is a predictable outcome of insufficient or erratic food provisioning. If the male is unable to provide a stable supply of food, the obvious explanation for delayed laying, the female is unlikely to remain faithful to the eggs. Hunger and a lack of vigor will force her to abandon the nesting effort in order to ensure her own survival.

But why do gyrs venture into this risky enterprise if the signals are there early that nesting success is unlikely? Why not abandon the nesting effort entirely? Why not defer production until another year when ptarmigan are more abundant and the prospects of nesting success are greater? The problem with this option is that the pair may lose possession of their nesting territory if they are away from the nesting site for extended periods. Being active at the site, but abstaining from breeding for one year gives gyrs a chance to try again the next year. Losing the nest site may foreclose any future options. Gyrfalcon loyalty to the nest site, even in years when ptarmigan were scarce, was apparent in the central Yukon. It was only after successive years of low ptarmigan numbers that any sites appeared abandoned, and those cases were few.

Why did the gyrfalcons not wait until water birds had returned and ground squirrels became active to begin breeding? I suspect the requirement for a long nesting period prevents this option. Such a delay would mean that young gyrs would leave the nest in the fall instead of mid-July, making survival much less likely. The genes that guided such a failed strategy would not be perpetuated.

Indeed, ptarmigan appeared to have a profound effect on gyrfalcon production. They influenced the number of gyrs that bred, the number of eggs they laid, the number of clutches incubated, the number of young that eventually left the nest, and the laying date.

To validate, or at least reinforce, my conclusions that were based on observations and inference, I supplemented food to two gyrfalcon nest sites in one year of the ptarmigan recession. I would shoot ptarmigan every morning and deliver one to each nest site the same day. The intent was that by providing gyrs with ample food in a year when ptarmigan were scarce, I could induce the pair to breed and produce young in a year when success was unlikely. I also predicted that if gyrs had ample food they would nest early; much earlier than would be typical in a year when ptarmigan were scarce, proving that ptarmigan are the driving feature in the reproductive ecology of gyrfalcons. But could I entice gyrfalcons to take food I provided? Many animals will not touch food that they did not kill themselves. Fortunately, gyrfalcons are more desperate. In Russia, it is well known that gyrfalcons feed on trapped ptarmigan (Pokrovsky and Lecompte 2011).

I began supplementing food at five gyrfalcon nest sites. At two nests, the pair would not touch the food I provided them; at another nest, it appeared the pair was infrequent or irregular tenants. The dead ptarmigan would freeze quickly in March if the pair did not use it immediately, making it difficult, perhaps impossible, for a gyrfalcon to feed on the carcass. One male and one pair, perhaps more desperate than the others or more loyal to the nest site, began taking the food I provided on a regular basis. The solitary male was soon joined by a female. I fed these two pairs each about forty-eight ptarmigan, almost 1.2 per pair per day, while they were courting and a few weeks into the incubation period. Now all I had to do was wait and see whether they would fledge youngsters.

I enjoyed my daily trips to the gyrfalcons' eyries to feed them. I would climb up a navigable part of the cliff with my crampon-type snowshoes onto a feeding perch. Here I would drop the ptarmigan, still pliable, and leave quickly, watching the tenants out of the corner of my eye, trying to avoid direct eye contact.

Gyrfalcons are particularly sensitive to intrusion or other disturbances during the courtship period. This crucial time sets the course of action for the events to come over the next four to five months. Knowing early whether the site is safe or not is critical. If the gyrs are displaced early, it gives them adequate time to switch to an alternate nest site and raise a brood. The birds I fed, surprisingly, tolerated me. They allowed me to approach within sixty meters, sometimes closer, while they were seemingly unperturbed and quiet. No display of aggression, distress, or the tell-tell "cacking" of agitated gyrfalcons. In fact, at one site the male barely gave me time to disembark from the feeding perch—a little knoll where much of the feeding and butchering was evident—before landing, clutching the ptarmigan, and parading with it in front of the female as if he had caught it himself, in typical courtship display.

The two gyrfalcon pairs that were fed did as I predicted. They each fledged three young, in a year when only 67% of all other pairs produced eggs, of which 24% of these nests were abandoned. The experimental pairs also laid eggs early—6 and 11 April—as compared to the other nests that did not benefit from extra food, where six of seven nests were not initiated until after 13 April and where the median laying date was 27 April. Indeed, the success at the two experimental nests compared well to the rate of production in the year when gyrs were at their best; when an average of 2.57 young were fledged from nests that were initiated around a median date of 8 April. Although a small sample size, I believe the results are convincing when considered with my other observations.

My observations and experiment bring me to the same conclusion. Gyrfalcon reproductive success depends on the abundance of ptarmigan. Remember that when gyrfalcons begin nesting, ptarmigan are the only prey available to them. In the central Yukon at the time of my experiment, the number of ptarmigan had crashed to their lowest point in the five-year study. But are there alternative hypothesis to explain gyrfalcon reproductive success?

DOES WEATHER MATTER?

Winter temperatures (average and minimum), snowfall, and snow accumulation, tested singly or together, had no apparent direct influence on the number of young gyrfalcons that were hatched or fledged during my study in the central Yukon. This finding is not surprising. Gyrs begin nesting in the winter, where even in an average year they endure brutal temperatures, wind chill, snow, and freezing rain. Gyrfalcons are well adapted to the northern climate, and they mitigate the impacts of weather by selecting nest sites that provide them some protection.

Although typical winter weather conditions did not appear to have a direct influence on gyrfalcon productivity in the central Yukon, perhaps unusual snow conditions indirectly affect gyrfalcon nesting success. In one year in the Mackenzie Mountains, as I mentioned earlier, there were surprisingly few gyrfalcons in the fall (suggesting that few young were produced) in a year when ptarmigan were common. However, the ptarmigan hatched young three weeks later than usual, likely due to persistent snow late in the spring. This took away the advantage that gyrfalcons normally have of vulnerable ptarmigan when gyrs begin breeding. Consequently, production was compromised.

At least two studies, one in the central Arctic of Canada (Poole and Bromley 1988) and one in Iceland (Nielsen 2011), found that spring weather is a significant factor influencing gyrfalcon reproduction. In both studies, it was proposed that cold temperatures and high precipitation in March and April may have affected gyrfalcon body reserves or influenced rock ptarmigan behavior.

I find it difficult to conceive that weather alone would have a direct effect on gyrfalcon reproduction. Severe weather is a common condition during the early part of the breeding season, and gyrfalcons have adapted to these circumstances, both physiologically and behaviorally. However, it seems plausible that snow conditions and perhaps severe temperatures could affect the availability and/or vulnerability of ptarmigan to gyrfalcons. There are a number of weather-related conditions that affect ptarmigan: delays in ptarmigan migration (Poole and Bromley 1988), postponement of laying (Hannon et al. 1988), availability of suitable cover (logically influenced by snow depth) (Gruys 1993), and possibly greater use of snow roosts in late winter, as influenced by the depth and conditions of snow and blizzard conditions. These weather-induced changes affecting ptarmigan are likely to handicap gyrfalcons. The male's hunting success will be compromised, and the female may be unable to gain enough weight to reproduce.

WHAT ABOUT GOLDEN EAGLES?

Golden eagles play an important role in the ecology of gyrfalcons, and they were common in the Ogilvie Mountains (three eagle nests for every one gyrfalcon nest). Aside from providing nest platforms, did eagles influence gyrfalcon reproductive success?

With ptarmigan scarcity accounting for most of the gyrfalcon failures, the additive effect of eagles was seemingly minimal. However, as I suggested earlier, gyrfalcon brood losses may have been higher when golden eagles were more active. It is conceivable that gyrs avoid nearby eagle nests at all costs and therefore may have to extend their search for prey farther from their home base, leaving their nests unguarded for longer periods, and ultimately suffering greater brood losses when eagles are common. Competition for food may also extend the gyrfalcon's hunting range (Johansen and Ostlyngen 2011). But in the Yukon, brood losses were small in comparison to failures due to low ptarmigan numbers.

Golden eagle with ptarmigan

EXPERIENCE OR HABITAT CHOICE?

I suspect experience and habitat choice give gyrfalcons a decided edge over potential competitors. If consecutive success at gyrfalcon nest sites is a measure of experience, I could find no apparent advantage of repeated success. Nest sites that produced young in one year were more apt to be occupied the next year, but there were no apparent benefits in egg laying, clutch size, laying date, or the number of young fledged, when compared to gyrfalcon reproductive outcomes from nests that were vacant the year before. Experience likely affects gyrs' success when contesting and claiming nest sites.

As for habitat choice, some territories probably host more ptarmigan than others, or support fewer golden eagles, and in these areas, I would expect better long-term reproductive success for gyrfalcons. However, I made no effort to measure the distribution and relative abundance of ptarmigan across the study area, so I had no hope of trying to figure out these details.

TO BREED OR NOT TO BREED – THE COURTSHIP STORY

If reproduction depends on the number or vulnerability of ptarmigan, and most adjustments are made early in the breeding season, how do gyrfalcons assess their prospects and respond accordingly? What role does courtship play, what behavioral cues are important, and how is this information communicated between mates? To understand the role of courtship better, I compared courtship events during a year when ptarmigan were abundant with a year when ptarmigan were scarce.

Courtship is a time when males and females are not only bonding, but communicating to one another information about the territory and about their own abilities. The female, in particular, benefits tremendously from knowing more about her partner. After all, she is completely dependent on him to provide food for her and her brood over at least a fifty-day period. Her success as a mother depends almost entirely on her choice in a mate.

If gyrfalcon reproduction is mostly about ptarmigan, then food deliveries to the female should be compromised during years when ptarmigan are less common. But I had not expected just how dramatic the difference would be. In the year that ptarmigan were abundant, I observed the male delivering a ptarmigan to the female on average every 21.8 hours. At one site alone, a male delivered three ptarmigan in 28.6 hours of observation, or on average, one every 9.5 hours. A year later, when ptarmigan had crashed, an average of only one ptarmigan was presented to the female every sixty hours—roughly 2.5 days between food drops!

Food is a great motivator. When there were long intervals between food deliveries, the females were less faithful to the nesting site. In good years, females were present at the sites I was watching 79% of the time. This attendance fell to 55% in the year when they received fewer food drops. Also, females visited the nest less frequently in the year of fewer ptarmigan. Apparently, she was less stimulated to begin laying eggs. This was further reinforced when I examined her loyalty to the site in relation to her eventual nesting success, irrespective of year. Females that succeeded in raising young spent almost 96% of their time at the cliff site during the courtship period; females that failed were around only 24% of the time during courtship.

Observation blind, Ogilvie Mountains

The male responded to the female's apparent apathy by increasing the frequency of his nest visits and displaying more aerial acrobatics. The male visited the nest on average every 2.9 hours in the year when food was scarce compared to every 4.2 hours when times were good. As I discussed earlier, a simple statistical regression indicated that the less time the female spent at the nest, the more the male visited the nest (72% of the day-to-day variation in male nest visits was influenced by the female's fidelity to the nest). His efforts appeared to be designed to entice the female to nest, and her lack of response motivated him to try harder. But alas, I suspect the female was simply responding to her physiological condition, tied to food deliveries.

Male acrobatic displays also varied with food supply. In years when ptarmigan were abundant, courtship seemed more subdued with fewer aerobatic flights and shorter duration of courtship displays. When ptarmigan were scarce, courtship appeared more spirited and urgent. During the year when there were fewer ptarmigan, I recorded over 170 male aerobatic displays, most of which (83%) were associated with nest visits, copulations, or female aerial displays. When ptarmigan were abundant, I observed only sixty male aerobatic flights, and only 58% of them were linked to other courtship behaviors. When tested across years, males from successful sites displayed in the air on average only every eleven hours while males from failed sites displayed on average every five hours. Apparently, the male responds to female indifference by spending greater effort trying to motivate her through acrobatic displays.

I think the gyrfalcon's solution to the question of whether to breed or not is simple and intuitive. The job of the male is to bring home food to the female, and eventually the family. She must gain enough weight and refrain from hunting in order to develop and incubate eggs. Her success in this enterprise depends almost entirely on whether the male is capable of providing for her. If she gains enough weight (a product of nutritional state), this will stimulate hormonal changes to induce courtship behavior and influence her physiological ability to lay eggs. Consequently, she will spend more time at the nest. If she does not get enough food she will be hungry, hormonal change will be inadequate, and she will be apathetic toward the nest. Eventually she will be obliged to hunt for herself, an activity that will further reduce her ability to gain weight and her desire to breed.

What did not make sense initially was why the male, prompted by the female's apparent apathy, didn't work harder to bring food to the nesting site, if food is what motivates the female. He spent only slightly less time at the site when ptarmigan were scarce—49% of the time versus 55%—and even stayed for longer durations at the site (on average 4.3 hours compared to 3.3 hours), yet he brought home much less food. Why not hunt longer, invest more time hunting and securing food, rather than linger at the nest site coaxing the female without the benefit of the gift?

There are perhaps two reasons to explain his loyalty to the nest site at the cost of fewer food deliveries. An obvious explanation is that he may run the risk of losing the female if he is away too long from the nest site. The female is a capable hunter herself and if he fails to bring home adequate food and is gone, she becomes hungry, has little reason to stay, and may be motivated to search for a better provider. If you remember the earlier analysis, the female's fidelity to the site was tied to whether the male stayed home in the preceding five-day period.

A second, perhaps less compelling reason for persistence at the nest site may be that if the male spends too long away from the site, he runs the risk that he will lose the area to a competing male. Failing to breed in one year, leaves options for him in future years, but only if he has possession of the nesting site.

Whatever the motivation, courtship events in the central Yukon varied from one year to the next and were instrumental in determining the outcome of breeding. Of no surprise, ptarmigan were the dominant factor influencing these events from year to year.

Female with her chick

GYRFALCON POPULATIONS – BOOM AND BUST

The ptarmigan cycle produces a boom and bust economy for gyrfalcons. In years when ptarmigan are abundant, a pulse of young gyrs enters the population with good prospects for survival, at least while ptarmigan remain numerous. However, in years when ptarmigan are scarce, few gyrs are produced and those that are hatched are apt to fledge late. The prospects of these late-fledged young are dismal. The floaters—adults without territories—are also likely to suffer high rates of mortality or disperse away from their natal area during the trough in the ptarmigan cycle (Potapov and Sale 2005). Similar dynamics have been observed in other northern predators that depend on cyclic prey, notably great horned owls (Rhoner 1997) and snowy owls (Schmidt et al. 2012).

Poor survival of gyrfalcons during the ptarmigan depression may have significant implications. It may be that there is a scarcity of potential breeders to occupy nesting territories when ptarmigan are abundant. This being the case, the number of gyrfalcon nesting pairs (and consequently production) would lag behind the peak in ptarmigan abundance.

Indeed, Olafur Nielsen (1986; 1999) found that gyrfalcon production in Iceland was not synchronized with the rock ptarmigan cycle, but rather gyrfalcon reproductive success peaked 2-3 years after the ptarmigan population crested. Nielsen observed incomplete occupation of gyrfalcon territories when the ptarmigan population was high, and suspected higher predation rates when ptarmigan were declining. But why was this pattern not evident in the Yukon?

A closer comparison of the data reveals that gyrfalcon reproductive success in Iceland (Nielsen 1999) underperformed Yukon gyrfalcons, but only when ptarmigan were at high density. When ptarmigan were at their peak, fewer Iceland gyrfalcon nest sites were occupied (34% versus 73%) and fewer territories fledged young (48% versus 75%) compared to Yukon gyrs, suggesting that Iceland gyrs were unable to capitalize on the surge in the number of ptarmigan.

At least two features distinguish Yukon gyrfalcon ecology from that of Iceland. In Iceland, there are no willow ptarmigan and no arctic ground squirrels. Yukon gyrs prey heavily on ground squirrels late in the summer and perhaps this improves the ability of young gyrfalcons to survive that critical period when they are attaining independence, as well as benefit the floaters. In addition, willow ptarmigan are more fecund than rock ptarmigan (Martin and Wilson 2011) so there should be more young willow ptarmigan, as compared to rock ptarmigan, during the recovery phase. This being the case, more Yukon gyrs should survive during the winter when the ptarmigan population is rebounding.

In Iceland, no arctic ground squirrels and proportionally fewer young ptarmigan may handicap the survival of young gyrfalcons and floaters, leading to a situation where there are insufficient gyrfalcons to fill vacant nesting territories when ptarmigan numbers are increasing. Consequently, we would expect gyrfalcon nesting success to lag behind the ptarmigan recovery.

Female feeding youngster, male standing behind

When ptarmigan populations crash, Yukon gyrs may be more severely impacted than Iceland gyrs. Willow ptarmigan typically suffer higher rates of mortality than do rock ptarmigan (Martin and Wilson 2011) so the crash may be more abrupt for willow ptarmigan. Also, willow ptarmigan may be less vulnerable to gyrfalcon predation than rock ptarmigan when they are at low densities. Willow ptarmigan avoid gyrfalcons by selecting habitats within high willows (Gurys 1993), of which there is an apparent surplus when ptarmigan densities are low, whereas rock ptarmigan are known to form large non-breeding flocks in the decline phase (Dave Mossop, pers. comm.). More young rock ptarmigan in large flocks may give Iceland gyrfalcons an advantage over Yukon gyrs during this part of the ptarmigan cycle. Accordingly, we should expect Iceland gyrfalcons to enjoy higher predation rates and therefore sustain their reproductive success as the prey population declines, as observed by Nielsen (1999).

Where gyrfalcons rely on willow ptarmigan, the dynamics are more comparable between populations. As I mentioned earlier, researchers in Sweden (Falkdalen et al. 2011) found that gyrs responded best when there were many young ptarmigan produced in the previous summer. Similar conclusions were drawn in Norway (Selas and Kalas 2007) where the number of breeding attempts by gyrfalcons was best explained by population indices of ptarmigan from the previous autumn. The effect is a one-year lag in gyrfalcon reproductive success to ptarmigan population size. I suspect the dynamics are similar in the Yukon. However, a delay was not as apparent in the central Yukon because the absolute number of young ptarmigan varied little in years when ptarmigan were abundant.

The consequence of population fluctuations may result in superior gyrfalcons. In poor ptarmigan years, those gyrs that fledge and survive are the likely offspring of talented, genetically superior parents. These skilled offspring come to occupy sites when years are good and so make a sizable contribution to the gene pool. These broods will be up against strong competitive pressure in subsequent years when food is scarce, further weeding out those that are not able to make the grade. Gyrs with the most vigor, then, are perpetuated. The Swedish team (Falkdalen et al. 2011) has provided some empirical support for this hypothesis. They found that 14% of all gyrfalcons they studied produced 50% of the offspring, leading them to conclude that high-quality territories and/or high-quality females made a disproportionate contribution to the gyrfalcon population. I suspect this general conclusion is also the case in the Yukon. However, I think premium males should be given credit for this success, as they presumably have a more demanding task: they are obliged to provide food to the females and the brood during much of nesting period.

CHAPTER EIGHT

THE SACRED FALCON

GYRFALCONS HAVE FOUND A HOME FOR THEMSELVES in one of the harshest and most desolate regions on the planet: the circumpolar zone in the northern hemisphere. Many are here not merely as summer residents but year-round as their Inuit name reveals—OKIOTAK, or literally, "the one who stays all winter." Two features here in the Arctic have had a profound hand in setting the evolutionary course for this avian predator: long brutal winters with limited daylight and a dependence on ptarmigan, the only obvious choice of prey in an otherwise stark environment. The outcome of this evolutionary challenge has been one of nature's most beautiful and exceptional creations.

The impressive physical traits and capabilities of the gyrfalcon did not go unnoticed by humans. Frederick II commented, "As a rule, all rapacious birds born [hatched] in the seventh clime and still further north are larger, stronger, more fearless, more beautiful, and swifter than southern species" (from Allsen 2006). The fact that the gyrfalcon stayed all winter in such a bleak environment enhanced peoples' admiration for this exceptional bird. Whether because of their beauty, talent, endurance, or rarity, we have long revered gyrfalcons. They have become prominent characters in our folklore across the north and beyond.

It is likely that the roots of this long-standing and remarkable association between gyrfalcons and humans began many years ago with ancient northern hunters, most of whom practiced some form of shamanism. The foundation of this belief is that all parts of the world are alive and interconnected, each possessing a spirit or invisible force. The spirits can be kind and generous, or hostile and vengeful. People with special powers—shamans, healers, medicine people, or prophets—are able to enter the world of these apparitions, usually for the purpose of restoring balance, healing, or appeasing the spirits. These medicine people would often use guides to help them enter the spirit world and interact with other entities. Gyrfalcons epitomized the virtues of a spirit-helper: strong, swift flyers, and remarkable hunters with particularly keen vision. According to Siberian shaman Sarangerel (2000), a powerful shaman received messages from the spirits of the Heavens through the gyrfalcon. The early spiritual role of gyrfalcons led to stories that became legends and cultural folklore around the northern hemisphere and even beyond.

I was first introduced to these ancient views through an Inuit friend of mine who was born on Baffin Island, where she grew up on the land. She shared with me stories of the days when the Inuit people would move from camp to camp to hunt, fish, and gather, in tune with the seasons. Some of these stories were about gyrfalcons. To see one, she told me, would portend good fortune.

Gyrfalcons were also a prominent object of Inuit folklore in the Western Arctic. In one story (told by Alunik et al. 2005), an Inuvialuit hunter, Kublualuk, acquires the power to transform into a gyrfalcon. It was a time of food scarcity. People were starving and in despair. Kublualuk ventured into the mountains south of Herschel Island in what is now the Yukon to search for game and seek guidance. But tired and hungry, he fell asleep on a mountaintop. The wind became strong and the skies cleared so that the moonlight lit up the mountains. Kublualuk awoke with the sound of the wind and the rustling of feathers from a large white gyrfalcon that was holding him in its talons. The gyrfalcon, sensing his fear and despair, told Kublualuk that he would teach him how to transform into a gyrfalcon so he could use this power to help his community during hard times. He then instructed the Inuvialuit hunter how to put on the gyrfalcon jacket to transform into a gyrfalcon, but reminded him to always think of the good of the people before he thought of himself. The large white gyrfalcon soared away into the sky and Kublualuk, now in the form of a gyrfalcon, scanned the land from above as never before. On his way home, he spotted a flock of ptarmigan and killed seven of them before transforming himself back to his human form. He walked home and shared his catch with his wife and the people from the community. For his part, Kublualuk became a great leader and is credited with the survival of the Inuvialuit during hard times.

In these ancient times in western North America, gyrfalcons were not only a spiritual icon, but were valued for their feathers as well. Their feathers were used by the Inuvialuit on arrows and spears, and were a special trade item (Irving 1953). Perhaps the functional worth of these gyrfalcon feathers was enhanced by their spiritual importance.

The fascination with gyrfalcons in Canada's high arctic reaches back at least into the Dorset culture (600 BCE – 200 CE) that predated the Inuit and Thule culture, as is evidenced by the discovery of amulets and carved spear shafts that depict gyrfalcons. Some of these carvings have "skeletal" lines incised on them and hollowed out inside, believed to represent the "un-perishable spirit of the soul" (Auger 2005). According to Robert McGhee (2001), the wearer of the amulets may have believed they would acquire hunting skills from the image of the gyrfalcon that was depicted. Similarly, the spear points embodied the virtues of gyrfalcons as spirit helpers. McGhee went on to suggest that, "The carvings of falcons or falcon spirits themselves must have had other significance, either as representation of a shaman's swift and far-reaching spirit helper or as containers of the spiritual powers required to successfully take animals as large and dangerous as walrus."

Taking Flight, sculpture by Abraham Anghik Ruben, depicting shaman emerging from a gyrfalcon displayed in the Kipling Gallery, Ontario

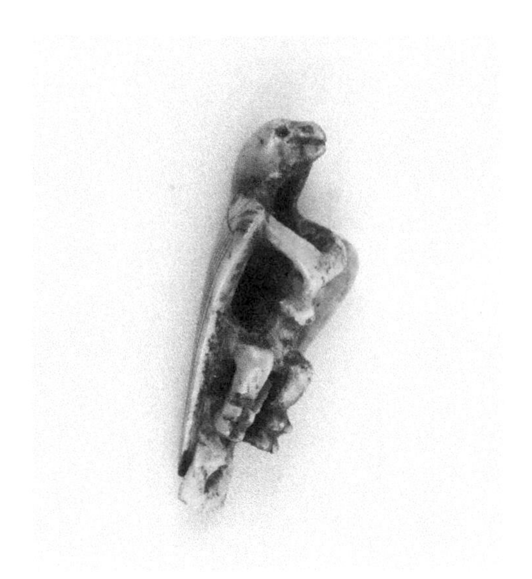

Dorset carving of gyrfalcon (side view); Dorset carving of gyrfalcon (top view)

The gyrfalcon also figures into Kaska Dena stories. The Kaska are another group of northern hunters who are part of the Na-Dene Athabaskan linguistic group and anchored to a vast area of boreal and subarctic mountains in Canada's northwest. Their history in North America, and their association with gyrfalcons, goes back many thousands of years to a time at least when the last major glaciers on the continent were receding. As noted, the Kaska have two words for gyrfalcons, one for the white gyr, and the other for the grey gyr. The white bird, claimed Charlie Dick, is synonymous with cold weather. He went on to point out that at one time, during cold periods, these white gyrs were common in Kaska lands. Today, however, they are rare and to see a white gyrfalcon foretells great fortune. The grey gyr, by comparison, is associated with warmer times and is more common today. The distribution of gyrfalcons across the north today corresponds with Kaska observations; white gyrs are more frequent where there is persistent snow and at the edge of ice-bound regions.

As well as being an omen of spiritual reverence and good luck, the gyrfalcon may also have been a spirit-helper or messenger to the Dena. One Dena elder referred to the group of birds we now refer to as falcons, as Ahda Dugudeji, or literally "she/he who tells the story of the Eagle." In Siberian folklore, the golden eagle was considered a symbolic ancestor of the Siberian people and the first shaman (Sarangerei 2000). Could it be that the gyrfalcon receives stories from the first great shaman? (It is widely believed by anthropologists that the Dena (Dene) are related to people from southern/central Siberia, based on language and cultural similarities (Vajda 2012), including similar folklore (Berezkin 2015).)

Adoration for the gyrfalcon was not confined to the North American continent. Legends about them have persisted amongst the people in Iceland, Scandinavia, Mongolia, Hungary, Russia, and even as far south as India.

The Norse culture in Scandinavia revered gyrfalcons through at least two significant gods, according to contemporary texts (Evensen 2007). The god Odin, the master of wisdom and war, was said to have been able to transform himself into a gyrfalcon to visit the Earth. Likewise, Freya, the goddess of love and fertility, was able to travel between earth and the underworld in the form of a gyrfalcon to return with prophesies (Mabie 1882). Popularized versions depict Freya dressed in a gown of gyrfalcon feathers. Whatever the manifestation, gyrfalcons were recognized as spirit-helpers which enabled the Norse gods to move between heaven and earth. A pictograph from a cave in northern Denmark, dated to be three or four thousand years old, shows a shaman transforming into a bird of prey (Busch 2015), most likely a gyrfalcon. This image was the inspiration for an exceptional soapstone carving by Abraha Anghik Ruben that depicts a shaman transforming into a gyrfalcon. The carving can be seen in the Kipling Gallery in Ontario.

Gyrfalcons are also part of Iceland's folklore (Sethi 2012). The legend of the gyrfalcon and the ptarmigan, although interpreted differently, conveys the interplay of mythology and ecology. The story is told that the gyrfalcon and the ptarmigan were close-knit siblings, who came from a single egg at the dawn of time and lived and played together. In one version of the story, their play one day got rough and the gyrfalcon accidentally killed her sister. When she realized what she had done, the gyrfalcon cried out in despair. Since then, gyrfalcons ceased to play with ptarmigan, and ptarmigan changed their feathers in tune with the seasons to hide from gyrfalcons. The memory of their companionship returns to the gyrfalcon in a symbolic way when she devours the heart of the ptarmigan. Even her call is considered one of anguish for having lost the kinship with ptarmigan.

Turul statute, Hungary

In what is current day Hungary, there are legends about a large mythical bird of prey called the Turul. The Turul was believed to be the messenger of God who sat at the top of the tree of life, below which perched other forms of birds who were believed to be the spirits of unborn children. In Magyar folklore, Turul is aligned with some well-known folk heroes: Attila the warrior Hun, Emese, the daughter of the great Sythian King Megog, and Almos and Arpad, the descendants of Emese who established the Magyar homeland of Hungary (Leviton 2012). But what is a Turul, and what are the roots of this mythology?

Most agree the Turul was a large dark falcon. Linguists claim that the word Turul has Turkic origins from the word Tugrul, which means gyrfalcon (Beni 2016). Quite likely, the Turul was what we now know as the Altai Gyrfalcon (Cade 1968). The Altai gyrfalcon – whether a pure gyrfalcon, a saker-gyrfalcon cross, or a subspecies of saker falcon, is a large, dark falcon found in the Altai region of central Asia where the breeding range of gyrfalcons may at one time have overlapped with that of saker falcons (Fox and Potapov 2001; Ellis et al. 2008). Like the gyrfalcon, the saker has all the elements of an iconic bird. It is large, elegant, rare, and formidable, and comes from the Altai Mountains, which according to Sarangerel (2000) was considered a sacred area. But how did this uncommon falcon, restricted to the Altai Mountains, become part of Hungarian mythology in the Caspian Basin?

There is much debate about where the Magyars (the Hungarians) originated (see Sandor 2013). Some have placed them in the southern Ural Mountains on the west Siberian Steppes, based primarily on language. Others argued they came from Asian stock, traced back to cultural similarities, including music and artifacts. Some legends indicate a link with Persia.

What is more likely, according to the extensive research of Frank Sandor (2013), is that the Magyars had their origins in the Indian subcontinent, settling near the Hindu Kush in what is now Pakistan and Turkestan, before emigrating north into southern Siberia east of the Ural Mountains. Eventually they were joined by emigrants from Persia. Frank Sandor argues that the Hungarian language likely had its roots as Sanskrit. As well as language similarities, he compares the near identical legends of the Hungarian Turul with its counterpart in Sanskrit, known as "The Hawk." Also, the early Hungarians were expert horsemen and horseback warriors, further pointing to a cultural link with the network of nomadic horsemen of the southern flanks of the Altai Mountains. Likely, these cultural similarities included the admiration of the Altai gyrfalcon—Turul. Perhaps there was a steady dispersal of nomadic people from the Asian steppe, a likely outcome of the emergence of the powerful Han dynasty in China in the first century CE and the realignment of many nomadic cultural groups in central Asia.

The Turul was celebrated by both the Huns and the Magyars. One legend claims that the Turul was the symbol of the great Hun Nation and the ancestor of Attila (Mena 2007). Attila the Hun, a military leader of great renown, was well known for his conquests against the Roman Empire (years 434-453) made possible with his exceptional cavalry of horse-mounted archers. Attila coveted the Altai gyrfalcon. It is believed that he had a shield on which was etched Turul (Cade 1968). In a museum in Hungary, there is a caricature of Attila. His breastplate bears the tree of life with the Turul at its apex. There is also a famous statue of Turul on the Buda Castle in Budapest, showing the large falcon delivering the divine sword to Attila.

The Turul is the centerpiece of at least two sagas that became pivotal in the development of the Magyar culture (Williams 1996). In the first of these stories, Emese had a dream in which Turul appeared and told her that her child would be the father of a line of great rulers of a great nation. In one variation of the legend, the Turul even impregnated her, venerating the pedigree. The key message was the foretelling of the great Magyar dynasty, with its origins from the house of Turul. Although her legendary son Almos never reached the Caspian basin (some say he was sacrificed to the gods for the cause of the Magyars), his son Arpad founded the Settlement of the Magyars in what is now Hungary in 895. As prophesied by Emese, this was the beginning of the famous Arpad bloodline.

In the second legend, a successor of Arpad and leader of the Magyar tribes had a dream in which eagles attacked their horses, but a Turul again appeared and drove the eagles away. This symbolic message prompted the Magyars to leave, thereby saving them from imminent defeat. The Turul then returned to the Magyars and led them to a new land, which would become Magyarorszag (the first Hungarian state) (Steves and Hewitt 2013). Arpad's great-great-grandson, Istvan (Stephen I), would become the first king of this Magyar nation in the year 1000 CE, and would expand Magyar control over much of the Caspian basin, further fulfilling Emese's dream of the gyrfalcon's prophesy.

Today there are a number of statues in Hungary that glorify Turul (Williams 1996): four large statues along the Liberty Bridge crossing the Danube, one in front of the Royal Castle of Buda, and one in the town of Tata in Northwest Hungary.

Throughout much of Central Asia, gyrfalcons were associated with war. War gods were an important feature in Central Asian shamanism (Nicolle 1990), and gyrfalcons epitomized the glorification of war. The Turkic people centered in the Altai Mountains honored their skilled warriors. Those able to shoot backwards as well as forwards were permitted to wear white gyrfalcon feathers in their helmets (Nicolle 1990). The Mongol Huns took the fascination further.

The rulers of the great Mongol dynasty, the Khans, who according to Sarangerel (2000) were considered the children and grandchildren of the spirits of the heavens, had a profound passion for gyrfalcons (Kahn and Cleaves 1999; de Rachewiltz 2004; Kolbas 2006). Genghis (also known as Chinggis), the great warrior who in the thirteenth century united the Turkic and Mongolian tribes across the Eurasian steppe into what became one of the most successful dynasties in history, was a powerful shaman who believed that the gyrfalcon was one of the lords of the sky. He believed that the gyrfalcon was the link between heaven and earth, a messenger relaying the will of the sky god Tengri (Fitzherbert 2006). Genghis Khan was even likened to a gyrfalcon (Cleaves 1982). To show his reverence, Genghis had coins minted that depict a gyrfalcon (Komaroff 2006). Genghis was particularly partial to white gyrfalcons. The color white was associated with good fortune, which was an essential component of Mongolian political ideology (Allsen 2011).

Genghis also believed that gyrfalcons were synonymous with military supremacy, and claimed that hunting with gyrfalcons (the practice of falconry) was the best training for warfare (Wingmasters 2001). Perhaps the alliance with heaven influenced the outcome of battle. Again, to show his admiration, Genghis adopted the gyrfalcon as his emblem, and made it the symbol of the great Mongol Empire, proudly depicted on the nation's banner (Secinski 2006). He also took gyrs into battle with him; history and art suggest he may have flown as many as five hundred white gyrs on his Asian military campaigns (Polo 1324). After one battle, Genghis, in complementing his warriors, was quoted, "You descended on your enemies like gyrfalcons from the sky" (Secinski 2006).

Genghis Khan Warrior, artwork by Vadim Gorbatov

Kublai Khan, Genghis's grandson and another powerful shaman who extended the empire through most of China, was similarly inspired by gyrfalcons. He introduced a gyrfalcon motif on tokens that were issued by the government to high-ranking officials to exchange confidential information through courier stations. The gyrfalcon token was particularly important. It was used for dispatching urgent military information. The high commanders of Kublai's great armies were also issued tablets with a gyrfalcon inscription to recognize their authority, claimed Marco Polo (1324): "To these dignitaries the Great Khan also gives a tablet with the sign of the gerfalcon; these tablets are given to the very great barons so that they may exercise full powers equivalent to his own."

In the epic thirteenth century chronicle of Genghis Khan written in the original Mongolian language a few decades after his death, titled *The Secret History of the Mongols*, gyrfalcons are again a prominent character (Cleaves 1982; Kahn and Cleaves 1999; de Rachewiltz 2004; Kolbas 2006). In this manuscript, gyrs are first mentioned preceding the birth of Genghis when his ancient ancestor, Boconcharchar, captured a gyrfalcon while living alone in the wilderness and taught this bird to hunt for him. The book also describes the matrimonial engagement of Genghis Khan. On his way to his wife's tribe in his search for a wife for his young son Genghis, Yisugei-Ba'atur met a man from another clan who shared with him a dream that he had the night before his unexpected visit. In his dream, a white gyrfalcon carried the sun and moon down in its talons and settled on his hand. This was taken as a sign that foretold the greatness of the Borjigin clan and the Mongolians. The dream and their arrival could not be coincidental, and so Yisugei decided that Genghis would marry the man's daughter, Borte.

Another story is recorded by Paulo Coelho (2006) in his book *Like the Flowing River*. This story, told to him by a Kazakhstan horseman, goes like this:

> *One morning, the Mongol warrior, Genghis Khan, and his court went out hunting. His companions carried bows and arrows, but Genghis Khan carried on his arm his favorite falcon, which was better and surer than any arrow, because it could fly into the skies and see everything that a human being could not.*
>
> *However, despite the group's enthusiastic efforts, they found nothing. Disappointed, Genghis Khan returned to the encampment and in order not to take out his frustration on his companions, he left the rest of the party and rode on alone. They had stayed in the forest for longer than expected, and Khan was desperately tired and thirsty. In the summer heat, all the streams had dried up, and he could find nothing to drink. Then, to his amazement, he saw a thread of water flowing from a rock just in front of him.*
>
> *"He removed the falcon from his arm, and took out the silver cup which he always carried with him. It was very slow to fill and just as he was about to raise it to his lips, the falcon flew up, plucked the cup from his hands, and dashed it to the ground.*

Genghis was furious, but then the falcon was his favourite, and perhaps it, too, was thirsty. He picked up the cup, cleaned off the dirt, and filled it again. When the cup was only half-empty this time, the falcon again attacked it, spilling the water.

Genghis Khan adored his bird, but he knew that he could not, under any circumstances, allow such disrespect; someone might be watching this scene from afar and, later on, would tell his warriors that the great conqueror was incapable of taming a mere bird.

This time, he drew his sword, picked up the cup and refilled it, keeping one eye on the stream and the other on the falcon. As soon as he had enough water in the cup and was ready to drink, the falcon again took flight and flew towards him. Khan with one thrust, pierced the bird's breast.

The thread of water, however, had dried up; but Khan determined how to find something to drink, climbed the rock in search of the spring. To his surprise, there really was a pool of water and, in the middle of it, dead, lay one of the most poisonous snakes in the region. If he had drunk the water, he too, would have died.

Khan returned to camp with the dead falcon in his arms. He ordered a gold figurine of the bird to be made and on one of the wings, he had engraved: 'Even when a friend does something you do not like, he continues to be your friend.' And on the other wing, he had these words engraved: 'Any action committed in anger is an action doomed to failure.'

Another well-known legend that honors gyrfalcons was staged in Muscovy in Russia in the early 1300s (Vaughan 1992; Potapov and Sale 2005). Triphon Patrikeev, a falconer of Tsar Ivan I, was flying the tsar's favorite white gyrfalcon, but it failed to return. The tsar ordered his falconer to find the falcon within three days or he would be executed by beheading, so Patrikeev went to Falcon Woods to find the gyrfalcon but without success. On the third day, exhausted in his search, he fell asleep. A dream came to him where a man appeared on a white horse with a white gyrfalcon on his right hand, and encouraged Patrikeev to look for the gyr on the tallest pine in the forest of Mytischtschi. He awakened and hastened to the area revealed to him in his dream, and there he found the gyrfalcon, which he promptly returned to Ivan. Patrikeev then repaid his gratitude by building a church at the site of his dream. Later the church was rebuilt and inside was decorated with a fresco that depicted Triphon on horseback with the gyrfalcon on his gloved hand. The artwork is now showcased in Moscow's Tretiakov Gallery (Potapov and Sale 2005).

Even as far south as India, gyrfalcons were revered in anecdotal accounts. According to legend, the ancient mother-goddess, Ka-dru, received from Shyena, or "frost-bird," the divine soma, or seed of life (Hewitt 1895). Shyena was a spiritual bird, able to fly between heaven and earth, and believed to be an eagle or falcon. A mantra in the Sanskrit Vedic sacred texts (Rig Veda IV.26.4; Griffiths 1890) described the iconic bird as follows: "Before all birds be ranked this bird, O Maruts; supreme of falcons, be this fleet-winged Shyena, because, strong-pinioned, with no chariot to bear him, he brought to Manu the god-loved oblation (Soma)." There seems little doubt that this fleet-winged, supreme of falcons, called the "frost-bird" referred to in this sacred text, was the gyrfalcon.

A more direct reference to gyrfalcons in India comes from artistic depictions and associated stories about Guru Gobind Singh and his white falcon. Falconry has had a long history in India, mentioned in Sanskrit literature and practiced well before the arrival of the Muslin dynasties (Taknet 2013). The last Guru of the Sikhs, Guru Gobind Singh, was particularly well known for his passion for falconry. He was often depicted with a large white falcon (presumably a gyrfalcon or gyrfalcon-saker cross) and referred to as Bazaanwale, "the one with falcon" (Kohli 2003).

Image of Guru Gobind Singh with falcon, artist unknown

Although not supported by historical accounts, there is an interesting story told in an article printed online (www.savethefalcons.org) that suggests that Gobind Singh's white falcon may have helped precipitate the decline of the mighty Mughal Empire (an Islamic Imperial power) on the Indian subcontinent. Resistance to the Mughals in the Punjab area led to a series of battles between the Sikhs and the Mughal in the late 1600s. At the time, messenger pigeons were used to deliver messages that were particularly useful during war to communicate battle plans. Falcons were used to intercept these military messages. According to the online story, at one point during one battle, the Sikh Guru Gobind Singh's famous white gyrfalcon caught one of the Mughal Emperor Aurangazeb's pigeons alive and instead of simply releasing the pigeon once the message was read, Guru Gobind Singh instead ordered that the message be altered before the pigeon was released. This set a trap for the Mughals and was a turning point in the battle. Soon after the Sikh Empire got a foothold in the Punjab and the Mughal Empire gradually lost its control over India. Whether or not there is merit to such anecdotal narratives, they do underline a widely-held admiration for gyrfalcons.

Stories of gyrfalcons also became part of Britain's folklore. During medieval times in Britain, it was a severe offense to possess a gyrfalcon. Only the royalty were entitled to own these birds. If caught, a commoner could be imprisoned, or have a hand cut off (Berners 1486, cited in Bartoslewicz 2012). There is a legend that a priest was wrongly accused of stealing a white gyrfalcon that belonged to the king. The priest was found guilty and sentenced to be hung, but as the priest stood ready for his execution, the lost gyr flew to the priest and perched on the gallows. The people of the town who were gathered to witness the execution believed this to be a sign of God and so set the priest free. The Pope heard about this miraculous event and decreed that the priest would be honored as Saint Albans, the patron saint of falconers, and that a church be built in memory of this event (Potapov and Sale 2005).

A similar story is told of St. Bavon, who was born of noble birth in what is now Belgium (Vaughan 1992). In his youth, he was accused of stealing a white gyrfalcon, and as with the plight of St. Albans, he was condemned to die. On the day of his execution the white gyr suddenly appeared, attesting to Bavon's innocence. He was freed and ordained as St. Bavon.

Gyrfalcons have a long history of veneration. Today they are the official bird of the Northwest Territories in Canada, they don the crest of the Icelandic Republic's coat of arms, and they are on the coat of arms of the Arkhangelsk Region of Russia. A white gyrfalcon was also made the national symbol of Kyrgyzstan when it gained independence from the Soviet Union in 1994. Some believe that the Ukrainian coat of arms, that shows a descending trident, is actually a stylized white gyrfalcon diving on its prey. The gyrfalcon is also honored on Iceland's highest decoration, the Order of the Falcon.

CHAPTER NINE

THE GYRFALCON AS A COMMODITY

FALCONRY IS A PARTNERSHIP BETWEEN HUMANS AND predatory birds, unique in that it is a bond with a non-domesticated wild animal. It involves, at least in the traditional practice, the controlled release of a bird of prey on a wild quarry. The sport was exemplified in times past by the pursuit of large prey: cranes, herons, bustards, swans, geese, and even wolves. This task of hunting large prey could be assigned primarily to gyrfalcons or golden eagles, the penultimate warriors. The pinnacle was the hunting of cranes with gyrfalcons, and the hunting of wolves on the Asian plateau with golden eagles. In these spectacular contests, involving prey that was much larger than the predator and where the prey would typically fight back, the outcomes were uncertain, adding a dimension of suspense. But why did this elaborate and time-consuming activity take hold?

THE ORIGINS OF FALCONRY

Before the dawn of civilization, the practice of falconry may have evolved from early hunters and gatherers to secure food and pelts, or to commute with spirits. The appeal to train raptors may also have been inspired by man's fascination with the ritual of combat, glorified by large birds of prey. Whether the motivation was practical or symbolic, human partnership with birds of prey has a long history. In this ancient story of falconry, gyrfalcons have become the centerpiece.

Most experts have speculated that falconry originated on the Eurasian Plateau or in the Middle East long before the advent of guns, where presumably a nomadic hunter (recall the Mongol legend of Boconchar) conceived of the notion of training a large bird of prey. Both regions were ideal for falconry—open, dry, and productive—but claims as to date of origin of falconry are without hard evidence.

Remains of birds of prey have been found among human remains in early settlements in the Middle East (Israel, Jordan, Syria, Iraq, and Iran) that date back 10,000 years (Dobney 2001). More recent excavations at Al-Maqar in Saudi Arabia have found falcon statues from about 9,000 years ago, together with evidence of the domestication of horses (Ghazanfar Ali Khan 2015). The vice president of antiquities and museums for the Saudi Commission for Tourism and National Heritage described these artifacts as "important symbols of authentic Arabian culture—equestrianism, falconry, the saluki hunting dog and wearing of the dagger" (Smith 2013). But all of this early evidence only confirms an association between birds of prey and humans, not necessarily a hunting relationship.

The literature citing old oral knowledge also places the origins of falconry back to the Neolithic Revolution when agricultural settlements became widespread about 8,000-10,000 years ago. However, these ancient oral accounts are difficult to authenticate. According to the International Association of Falconry and Conservation of Birds of Prey (2005), historical accounts from Iran credit Tahmooreth, a king of the Pishdadid dynasty that ruled about 10,000 years ago, as the first to have used birds of prey for hunting. In Iraq, it has been proposed that falconry was widely practiced 3,500 years ago, with the first record from the Epic of Gilgamesh written 2,000 years ago (International Association of Falconry and Conservation of Birds of Prey. 2005). Greek historian Ctesias of Cnidus mentions a mysterious race of people in Central Asia called "the Pygmies" who hunted hare and fox with birds of prey and ravens about 2,400 years ago (Aegisson 2015). In China, the written records trace falconry to about 3,500 years ago, according to Xiaodi and associates (2001).

The earliest physical evidence of falconry, according to Canby (2002), comes from an illustration dated at least 5,000 years ago, found at the Tell Chuera site in Syria in an area of ancient Anatolia where falconry has a long tradition. The drawing depicts "what appears to be a falconer holding up his dead prey" (Canby 2002). Another piece of evidence was unearthed in Turkey and is estimated to be 3,500 years ago. It shows a large bird of prey holding a hare, sitting on the fist of a human figure (Dobney 2001).

Dobney (2001) poses an interesting theory that indeed may tie falconry back to the Neolithic Revolution as a precursor to the domestication of animals and crop farming. He pins his theory on the mix of animal bones in the archaeological evidence as well as the plausibility of training birds of prey (and other birds) in ancient times. The early archaeological evidence in the Middle East 10,000 years ago suggests a major shift in human diet toward small animals. It appears large hoofed animals were replaced with fox, hare, partridge, francolins, and sandgrouse. Dobney proposed that a decline in their primary prey may have inspired early human hunters to find ways to shift their diet toward smaller animals, using a variety of innovative tools and methods.

Another plausible explanation for the shift toward smaller prey and the introduction of falconry may simply be tied to the introduction of crop farming. The move toward an agrarian lifestyle would have favored higher population density, a more sedentary lifestyle, and perhaps more free time (at least after the crops were harvested). A shift toward smaller more fecund prey would be expected when more people occupy smaller areas—fewer encounters with wide-ranging large mammals, and greater local hunting pressure. More free time may also have allowed hunters to experiment with alternative hunting strategies, such as the use of hawks and falcons. It is well known that falconry requires extensive training, takes much time to perfect, and requires special facilities (Hardaswick and Christopher 2011). A sedentary existence and free time, possible in a crop farming society, may have provided the time and circumstances to practice the sport.

But perhaps there is a simpler explanation. Many birds of prey, when taken from the nest, will form bonds with their captors if adequately fed and housed. Bob Collins suggested that the obvious next step in this relationship is to train these hunting companions to capture prey in return for food. Could it be that the origins of falconry are the logical outcome of this innocent relationship between a young boy taking a bird for a pet, which was eventually transformed into a hunting partnership?

Aegisson (2015) offered an equally simple and compelling theory for the birth of falconry. He suggested scavenging by humans for food at known nest sites by taking a share of prey brought to the chicks may have led to the eventual capture and training of birds of prey to capture and provide food for humans.

There is also some speculation that falconry may have advanced to help Neolithic hunters kill large mammals. It is surmised that birds of prey could have been used to confuse and distract large animals such as asses, wild horses, and pigs so that human hunters could more easily approach and kill them. Stephen Bodio (Cutchins and Elliason 2009) describes a petroglyph found in Tamgaly Gorge, Kazakhstan, etched into the rock 4,000–6,000 years ago. One image shows what appears to be a wild ass assaulted by two ring-tailed dogs (one grabbing the nose of the wild ass). On the back of the ass is a crudely defined figure. Above the ass is an image of a horse with what looks to be a saddle, and behind the horse is a clearly defined human figure. Bodio claims. "Every falconer in two cultures to whom I have shown this photo [of the petroglyph] identifies the figure [the one on the back of the ass] as a bird, facing the observer with wings spread. Most think it is an eagle, and a few believe it to be a Saker falcon..."

Whether the impetus was to secure small game or to help in the hunting of large mammals, falconry is thought to be rooted in the Middle East or the Eurasian Plateau. Perhaps two cultural paths came together by way of war and trade to employ falconry and venerate large raptors. Whether the idea of falconry spread westward across Asia and into the Middle East or arose independently within these regions, the one common feature is that it was established in open landscapes—grassland steppes, plains and deserts. Perhaps the hunting of wildlife by humans was difficult in these open landscapes without the aid of birds and dogs. Also, it is here in these open habitats that falcons are most at home, unencumbered by brush and forest cover, and are able to be tracked by the falconer with relative ease.

Perhaps horses, an important feature of these open grasslands and to the people that resided in these areas, were essential in these wide-ranging hunts. There seems to be increasing speculation that the domestication of horses and dogs coevolved with the use of birds of prey. Stephen Bodio (www.humansandnature.org) in his interpretation of the petroglyph in Kazakhstan, added, "This is the first picture of a local hunting team whose cooperation is engineered by humans," reminding us that "the hunters in Kazakhstan and the neighboring countries still use eagles with ring-tailed Saluki-like dogs and horses."

Saker With Kill. Tamgaly Gorge, Kazakhstan, artwork by Vadim Gorbatov

THE SPREAD OF FALCONRY – INVASIONS, ALLIANCES, AND TRADE NETWORKS

Falconry and the pre-eminence of gyrfalcons may not have taken hold and flourished across Asia and Europe were it not for the emergence of two centralized super-powers, the disruption and dispersal of small nomadic tribes, and the trading routes that developed. Quite likely climate change had a hand in setting the course of these events.

The first century BCE was an unstable period on the Eurasian Steppe. It was a time when many militant tribes mounted on horseback were struggling for land and power. According to Bas van Geel and associates (2005), these inter-tribal tensions were a result of population growth and prosperity beginning in the first millennium BCE, made possible with the increasing humidity on the Eurasian plateau that yielded greater forage production for livestock. The Yuezhi tribes (a sub-group of the ancient Scythians) centered in the Tarim Basin south of the Tien Shang Mountains, became the predominant power on the steppes, but they were constantly under attack from the Xiang-nu tribes (the ancient Huns), who eventually defeated them (O'chee 2008). The Yuezhi retreated west to what is now Tajikistan and Uzbekistan while the Xiang-nu Empire continued to grow, expanding south and east into China.

The Chinese under the Qin dynasty were unable to mount an effective defense against the invading Huns whose cavalry proved effective against the Chinese infantry (Wood 2002). In response, the Chinese sought resources and developed technology to overcome the mounted warriors (McLaughlin 2016).

According to the Selections from the Han Narrative Histories (www.dept.washington.edu), the Chinese also pursued an alliance with their western neighbors, the Yuezhi, in order to initiate a common assault on their northern adversaries. Chinese martial emperor, Wu-Ti (140-87 BCE), sent his official Chang Ch'ien and an envoy westward to initiate such an alliance. On his mission, Chang Ch'ien learned about rich kingdoms to the south and west, including a place called Li-Jian, which scholars believe was the Greek civilization, and the distant land occupied by the Romans called Da-Qin or "Great China" (McLaughlin 2016). Of more immediate interest, Chang Ch'ien was introduced to a remarkable breed of horse known for its endurance and strength—the famous Heavenly or Celestial Horses of the Ferghana valley. The virtues of this horse breed were not lost on the Chinese, who had given up much ground at the hands of the superb cavalry of the horse-mounted Huns. After many military excursions into the Ferghana valley, the Yuezhi (who initially were unwilling to align with the Chinese against the Xiang-nu tribes) eventually surrendered to the Chinese. With the help of his newly acquired horses, Wu-Ti was able to defeat the Xiang-nu in a series of battles around 129-121 BCE, and began consolidating control over the western and northern regions.

Chang Ch'ien's exploratory trail soon became a military road to Central Asia and soon after part of a vast trade network between the east and west (Hopkins 1980). Horses from the west and silk from the east were the foundations of this commerce (Wood 2002), hence the Silk Road. But trade rapidly radiated across Asia and into Europe and North Africa. Other goods entered the market place, including rhinoceros horn, tortoise shells, and many different foodstuffs (Wood 2002), as well as steel, furs, cotton, and alfalfa (McLaughlin 2016). Exotic animals were also exchanged, including falcons (Haksoz and Usar 2011). One can assume that gyrfalcons, given their prestige and high value, were part of this commercial trade. Notwithstanding the trade in products, there was also the exchange of technologies. It is probable that those falcon-hunting skills practiced by the nomadic people on the Asian steppe, and those from the Persians and the Arabs on the deserts, were exchanged, adapted, and perfected to suit local circumstances.

Concurrently, small independent tribes were pushed west with the expansion of the great empire of China during the Han Dynasty and the gradual decline of the Roman Empire (Wells 1922). No doubt, they took their falcons with them. Attila the Hun, who united many of these nomadic tribes and unleashed his terror on the Roman Empire in the fifth century (Davis 1999), and who coveted the Turul falcon (de Grubernatis 1872, cited in Skuker 2015), quite possibly had a hand in introducing falconry into Eastern Europe. This vast network of trading routes (Haksoz and Usar 2011), together with the westward spread of the Huns made possible an emerging interest in falconry across Europe. Gyrfalcons were at the top of the queue—the falcon par excellence and the object of special expeditions.

FALCONRY IN EUROPE

The practice of falconry in Europe was perhaps first recorded by Pliny the Elder (77 CE), who wrote, "In the district of Thrace inland from Amphipolis men and hawks have a sort of partnership for fowling: the men put up the birds from the woods and reed-bogs and the hawks flying overhead drive them down again; the fowlers share the bag with the hawks." There are also suggestions that Julius Caesar may have used falcons to destroy pigeons carrying messages in about 100-44 BCE (Aegisson 2015).

The first written reference to falconry in Europe, according to the scholarly research of Robin Oggins (2004), is a manuscript written in France in year 459 called Eucharisticus of Paulinus of Pella. About the same time, depictions of birds of prey and falconry started showing up in sculpture and on coins. Aristocratic graves in Sweden, dated between 500 and 1,000 CE, frequently contained the bones of gyrfalcons and other birds of prey not found in earlier graves (Ericson and Tyrberg 2004). There are also records of falconry being regulated in the Salic Laws from the reign of Clovis I, the first king of the Franks, around 500 CE (Jennbert 2011), as well as in Burgundian Laws at the same time (Aegisson 2015). Tommy Tyrberg (2002) suggested that falconry was introduced into Sweden through contact with the nobility of the Germanic states after the collapse of the Roman Empire. Presumably, these Germanic people encountered the tradition of falconry from their second century forays onto the Eurasian Plateau or from the Huns who invaded them in the fourth century. Falconry was thought to have been introduced to Anglo-Saxon England in the late sixth or early seventh century, similarly based on the preponderance of raptor bones associated with human graves, depiction of birds of prey in art, and the occurrence of small copper-alloy bells (Wallis 2017).

The Russians were likely exposed to the sport from contact with the oriental people on the Asian steppe at the same time that the sport was introduced to other parts of Europe. However, the written records do not place falconry in Russia until about the ninth century with the establishment of a falcon-house in Kiev by its first ruler, Oleg (Kovalev 2012). By the thirteenth century, it was common knowledge in Europe and the Middle East, that the best hawking birds were found in the Slavic, Prussian, and Russian countries. Gyrfalcons, which topped the list of important hawking birds, were thought to be abundant in Russia (Allsen 2006). At the time gyrfalcons were sought after in Russia, they were also being captured and traded from Iceland, Scandinavia, and Greenland.

FALCONS, FEUDALISM AND CLIMATE CHANGE

Europe was ripe for the spread of falconry around the sixth century CE. It was a time when three significant events coincided: the collapse of the Roman Empire, the emergence of feudalism, and the warming of the climate across Europe and the Atlantic.

With the decline of the Roman Empire in Western Europe (about 500 CE), a political vacuum was created. The breakdown of a strong central government led to the emergence of small militaristic groups across Europe. Land tenure was unstable, the economy was crumbling, there was no longer a common currency, and there was a widespread collapse of trade (Perry 2013). Political chaos led to violence. A warrior society took hold, with feudal landlords offering protection to their servants (vassals) in exchange for military services. With this feudal society, social classes were ranked depending on their prosperity, further dividing citizens by wealth and influence. Falconry became the badge of affluence.

Falconry blossomed further during the Crusades. The Crusades were religious-based military campaigns intended to free the Christian holy lands from Muslim power. There were a series of Crusades, initially by the Knights of France to free Jerusalem in 1095 and continuing into the fifteenth century (Asbridge 2004). Crusaders brought home with them many of the devices and jargon of eastern falconry.

One particularly enthusiastic falconer of this era was Frederick II of Hohenstaufen, Holy Roman Emperor and King of Sicily and Jerusalem. Frederick was notable in that he was "one of the two baptized sultans of Sicily" (Shahid 2009; Haskins 1922). It has been suggested that he returned from the crusades in 1228 with experienced falconers from Arabia and Syria and some of the contrivances of falconry (including the hood), furthering his expertise in falcons and falconry. According to Philip Hitti (2002, cited in Shahid 2009), Frederick brought skilled falconers from Syria and had his interpreter-astrologer translate a renowned Arabic treatise of falconry (believed to have been written in the sixth century by al-Ghitrif, the master of the hunt for the caliphs in the Middle East). This translation is thought to have become the basis of Frederick's work on falconry and natural history: *De arte venandi cum avibus* (The art of hunting with birds, circa 1240). Gyrfalcons figure prominently is this work. [See Hardaswick and Christopher (2011) for contemporary falconers' high evaluation of Frederick II.]

Samurai With White Goshawk, artwork by Vadim Gorbatov

At the same time that feudalism was spreading across Europe, the climate was warming (Lamb 1966; Fagan 2008). Europe was experiencing the Medieval Warm Period (800 – 1300 CE). Food was plentiful and trade was re-established. The population grew, more lands were cleared and cultivated, and landlords became even more powerful. These were times of affluence and opulence for the aristocracy, staged in a setting of castles and manors, controlled by kings and feudal lords and defended by their knights.

Falconry became particularly extravagant (see Oggins 2004). The nobility became preoccupied with this sport as the class system flourished. Expensive expeditions were undertaken to obtain falcons and the trade in the birds became a lucrative business. Falcons were used as accessories to exaggerate wealth and social standing. They accompanied aristocrats to church and banquets and on excursions, trade missions, and to war.

GYRFALCONS: THE ULTIMATE PRIZE

Gyrfalcons, although not the most commonly used raptor in the sport of falconry, were the most cherished. As well as being rare and large, gyrfalcons have desirable characteristics. Frederick II considered the gyr to be the best bird for hunting "out of respect for their size, strength, audacity, and swiftness" (Wood and Fyfe 1943). Likewise, Daude de Pradas (1214-1282), a French troubadour who wrote his own treatise on falconry titled *Augels Cassadres*, considered the gyrfalcon to be the fastest and the most resourceful bird of its size (Oggins 2004). Added to their remarkable capabilities as a hunter, gyrfalcons had "the perfect nature of the falcon in appearance, color, action, and voice," wrote Albertus Magnus, the famous thirteenth century German philosopher, theologian, and author (cited in Oggins 2004). Genghis Khan also recognized the virtues of gyrfalcons. According to Persian sources, "When Genghis Khan asked his officers what is the greatest pleasure of man, they all answered that hunting with gyrfalcons is the supreme joy" (Allsen 2006).

The gyrfalcons' esteem may have also been romanticized by the mystique of its northern home. As Allsen (2006) stated, "The fact that these birds came from the 'land of darkness,' an alien and mysterious clime with extreme conditions of temperature and light, 'naturally' endowed gyrfalcons with very special properties in the minds of southern falconers."

The gyrfalcon was particularly valued for its ability, when trained, to hunt large prey such as common cranes and gray herons. These birds are much bigger than gyrfalcons, and known to fight back. The crane will defend itself with feet, beak, and talons, and with help from other cranes when they are in flocks (Wood and Fyfe 1943). The heron's defense is its lethal beak. C.B. Carpenter described in Patterson's (1879) report to the Belfast Natural History and Philosophical Society, in their proceedings of 1878-79, "The Heron can so turn its neck as to cause its bayonet-like beak to project upwards behind the wing

Atque palam est, quod ars venandi cum avibus & ars est, & ideo prior.
& cæteris venationibus nobilior & dignior,
De Arte Venandi cum Avibus
FRIDERICI II. Lib. I. Cap. 1. 1248

Previous page: De Arte Venandi Cum Avibus of the Emperor Frederic II, artwork by Vadim Gorbatov]

at the same time that its flight is continued, so that, when the falcon darts downwards, it runs the risk of being impaled upon this formidable weapon."

Although gyrfalcons in the wild do not eagerly pursue large quarry, some will pursue large prey without hesitation when rigorously "trained-up." Guillelmus Falconarius (William the Falconer), one of Italy's legendary falconers and an author of one of the three significant manuscripts on falconry during the Middle Ages (Shailor 1991), claimed that the gyr is "the one who fought best against the large birds such as cranes, herons, and geese." Or as Frederick the II noted, "She [the gyrfalcon] is very easily taught to hunt everything that any other falcon can chase, and with greater facility and swiftness since she excels in courage, power and speed.... the crane falcon par excellence" (Woods and Fyfe 1943). Pertti Koskimies (2011) paraphrased the great Nordic historian, Olaus Magnus, who said, "The gyrfalcon is strong and furious enough to rush to hunt up to five common cranes and not to stop until it has killed all of them." In addition to the impressive spectacle of these contests, cranes and herons were considered royal delicacies, often on the royal menu at Christmas and other festive occasions, according to Robbin Oggins (2004).

But gyrfalcons are difficult to maintain in good health. They are susceptible to some of the parasites and diseases to which they have not become adapted and which are typically common in southern latitudes where gyrfalcons were held. Dementiev (1960) noted, "The northern gyrfalcons do not stand the Central-European climate very well, and the majority of them die for reason of various diseases." Allsen (2006) also commented that despite special transportation protocols in Russia, there was a high mortality of gyrfalcons in transit to distant lands. This would have meant even fewer gyrfalcons to trade, further escalating their value.

Being both rare and desirable, gyrfalcons commanded a high price, such that only the wealthy, generally the nobility, were able to afford such birds. In England where there are written records of transactions, royal audits have found that four of the five most expensive purchases for birds of prey were for gyrfalcons. The highest price recorded for a gyr was roughly equivalent to 40% of a knight's annual salary, says Oggins (2004). In Iceland, the price for twelve gyrfalcons was equivalent to approximately fifty tons of cereal grains—enough grain to meet the necessities of life for 100-200 people for a year (Klaus Friedland 1964, cited in Aegisson 2015). In the Arab states, their value was on par with purebred Arabian stallions, and in Georgia, their worth was equal to that of the much-desired Argamak horse (Potapov and Sale 2005). As John Fryer (1689) commented from his travels within Iran in the 1670s, the "hawkes of Moscovia [gyrfalcons] command a great price among the Persian nobles" such that "they seldom appear abroad without one of them on their hand." Indeed, the zeal for gyrfalcons was widespread in Eurasia, as noted by Allsen (2006), who remarked that there was a "mad passion for Falco rusticolus, a passion that diffused across the continent."

The trade in gyrfalcons likely took hold in the east and spread westward (Allsen 2006). Allsen claims it began with the Qitan Mongols, who founded the Liao Dynasty in northern China and Manchuria in the first century. Their quest for gyrfalcons took them to the Amur region in Siberia. Here they obtained and exchanged gyrfalcons through intertribal alliances and an elaborate tribute system. Eventually this traffic developed into an extensive trade across Eurasia.

To safeguard its supply and status, the nobility in most countries made it illegal for the lower classes to fly gyrfalcons (Potapov and Sale 2005). There were often harsh penalties for those of undeserving class who were caught poaching young falcons from the nest or in possession of the noble birds. In some society's it meant death, in others the mere removal of a hand or arm, or imprisonment. Juliana Berners, in her fifteenth century publication, *The Boke of St. Albans*, provided a symbolic alignment of falconry birds with social classes, and decreed that the gyrfalcon was to sit as the bird of the kings (Berners 1486). Although it is well known that gyrfalcons were not solely flown by kings, they were associated with the nobility in Europe. They were the most privileged and consequently could afford to have gyrfalcons. In Norway, Denmark, Iceland, France, the Netherlands, and Russia, the royalty, at times, had a monopoly over the capture and acquisition of gyrfalcons (Seaver 1996; Potapov and Sale 2005; Shrubb 2013).

Recall that the great Khans, notably Genghis and his grandson Kublai, had an intense interest in gyrfalcons. Marco Polo (1324), the famous Italian adventurer, claimed that Kublai had at least 1,000 gyrs; 200 alone at his summer palace at Xandu in inner Mongolia, where he would visit them at least weekly to feed them and admire them. Marco Polo described the Great Khan's passion for gyrfalcons like this:

When the Great Khan goes on the journey of which I have told you towards the Ocean, the expedition is marked by many fine displays of huntsmanship and falconry. Indeed, there is no sport in the world to compare with it. He always rides on the back of four elephants, in a very handsome shelter of wood, covered inside with cloth of beaten gold and outside with lion-skins. Here he always keeps twelve gerfalcons of the best he possesses and is attended by several barons to entertain him and keep him company. When he is traveling in this shelter on the elephants, and other barons who are riding in his train call out, 'Sire, there are cranes passing,' and he orders the roof of the shelter to be thrown open and so sees the cranes, he bids his attendants fetch such gerfalcons as he may choose and lets them fly. And often, the gerfalcons take the cranes in full view while the Great Khan remains all the while on his couch. And this affords him great sport and recreation. Meanwhile the other barons and knights ride all around him. And you may rest assured that there never was, and I do not believe there ever will be, any man who can enjoy such sport and recreation in this world as he does, or has such facilities for doing so.

Previous page: The Kublai Khan Hunting Trip, artwork by Vadim Gorbatov

GYRFALCONS AND THE CRUSADES

On more than one occasion, gyrfalcons were distinguished during the Crusades. In one particular campaign in September 1396, gyrfalcons gained considerable notoriety (Capainolo and Butler 2010). In the long-standing fight to restore Christian control over the Turkish controlled Holy Lands, an allied force of Crusaders organized by the King of Hungary and representing many parts of west-central Europe invaded the Turkish controlled fortress of Nicopolis, now in Bulgaria. Part of the crusaders' contingent included knights from the medieval French Kingdom of Valois-Burgundy. An initial cavalry charge by the French brought them into a trap laid by the Ottoman Turks, under Sultan Bajazet. Most of the French forces were killed or captured and the crusaders were defeated. The Duke of Valois-Burgundy, Phillip II (the Bold), despairing the capture of his son, negotiated for his freedom, along with twenty-four other knights, for the price of 200,000 gold ducats. But the Sultan wanted something even more precious, and so a deal was sealed for the release of the captives at a price of twelve white gyrfalcons.

GYRFALCONS AS TRIBUTES

Gyrfalcons were also important as gifts of state, or offerings of peace, throughout Europe and Asia (Allsen 2006). They served as tributes that were part of a system that demanded obedience through the ritual exchange of gifts. Gyrs were also offered to allies and neutral states in exchange for support.

The tribute system may have begun In China. At least as far back as 907, the Liao court required its northern neighbors, the Jurchen, to levy a tax upon the tribes of the Amur River to supply gyrfalcons. To accomplish this, Liao officials commanded the Jurchen to mount major military campaigns sometimes involving one thousand horsemen to secure the tribute of gyrfalcons (Allsen 2006). This oppressive duty sparked the eventual rebellion and overthrow of the Liao by the Jurchens in 1125 (Twitchett and Tietze 1994). The Mongols of the Yuan instituted a similar system designed to secure gyrfalcons, according to Marco Polo (1324). The Ming Dynasty, which established control of China in 1411, also demanded an annual tribute of gyrfalcons from the northern Jurchen tribes (Allsen 2006). Indeed, across many Chinese dynasties gyrfalcons were considered "a diplomatic tool, to be monopolized and utilized for political ends" (Allsen 2006).

The diplomatic and commercial exchange of gyrfalcons was also evident in the west. As far back as the early 900s, Hakon of Norway was known to have sent fifty gyrfalcons to King Harald of Denmark to maintain diplomatic favor (Bo 1962, cited in Potapov and Sale 2005). From the twelfth century onward, there are a number of references of gyrfalcons being bought, sold, and exchanged as gifts (see Allsen 2006, Vaughan 1992, Oggins 2004). One record "renders account for one hundred Norway Hawks of which four are to be white" (Jennbert 2011). Another has King Haakon IV of Denmark in 1224 sending the English King Henry III 13 gyrfalcons, together with many walrus tusks, in an effort to forge a trade alliance (Oggins 2004). He also sent gyrfalcons to the Sultan of Tunis and Frederick the II in Sicily (Aegisson 2015). After the thirteenth century, the records indicate that there was lucrative trade in falcons between Scandinavia, the Baltics, Western Europe, the Balkans, the Middle East, and North Africa (Jennbert 2011). In one record, eighty gyrs were captured near the Arctic Circle and sent to the Sultan of Babylon for his crane hawking. The importance of gyrs even extended into the Indian subcontinent. In 1675, Russian officials were told by Indian merchants that, "it would be best of all and most gratifying to Indian Rulers [if the Tsar] allows to be sent gyrfalcons and hawks" (Dale 1994).

Gyrfalcons were valued as tributes in Russia perhaps more so than in any other European country (Allsen 2006). Oppressors received gyrfalcons, but as well, militant neighbors were presented with gyrfalcons to diminish their inclination to attack, and neutral states were given gyrs to buy their support. Gyrs were also gifted to smooth over misunderstandings. Indeed, gyrfalcons were crucial to Russia's self-preservation (Potapov and Sale 2005). This was particularly so during the time when the great Khans of Mongolia were gaining prominence in Eurasia. In 1221, Genghis Khan invaded Russia at Kalak River to establish a foothold (Grant 2011) and later (1236) Moscow was attacked, plundered, and burned by a wing of the Mongolian Empire (Nicolle and Shpakovsky 2001). Although the city recovered, it became a protectorate of the Mongols. To appease the Khans, it was well known among their Russian conquests that the Mongol leaders favored "gyrfalcons, furs, women, and Kyrgyz horses" (Alikuzai 2013). Taxes and gifts (notably gyrs) were sent to the Mongol rulers preventing further repercussions. It soon became common knowledge that Russian gyrfalcons were the best hawking birds and were abundant. Marco Polo in 1295 comments on Russia and its wealth of gyrfalcons: "The country is so great that it reaches even to the shores of the Ocean Sea, and 'tis in that sea that there are certain islands in which are produced numbers of gerfalcons and peregrine falcons, which are carried in many directions."

As well as the rulers of the Mongolian Empire, the Sultans of the Islamic Empire also had an obsession for Russian gyrfalcons, notably the white ones, for sporting but also as diplomatic presents. The Persians, in particular, considered Russian gyrfalcons to be the greatest of presents. Dementiev (1960) claimed that among these gifts to Persia there had to be a special gyrfalcon—a white gyr with accessories (bells, jesses, hood and even a ceremonial breast and tail plates) inset with gold and precious stones.

In his prose novel, *Felix*, or Book of Wonders, published in 1288, Raymond Llull mentions "many men with gyrfalcons, which they had brought from one end of the world and were now taking to the Tartars to make money" (cited in Allsen 2006). Afanasii Nikitin, a merchant of Tver (Russian vassal state) on route to India in 1475, also mentions the exchange of gyrfalcons in his journal (see Allsen 2006): "And I waited two weeks in Novgorod for Hassan-Beg, the Tatar ambassador of Shirvan (now Azerbaijan). He was traveling with gyrfalcons from the Grand Duke Ivan, and he had 90 gyrfalcons with him." Allsen (2006) remarked, "Indeed the Russians often played their "gyrfalcon card" in the course of contentious relationships with the Muslim states to the south. Gyrs were so valuable that their export became a monopoly of the state and they were put on the Russian list of prohibited goods, and "could not be exported or even transported within the realm on pain of death" (Allsen 2006).

THE SOURCE OF GYRFALCONS

Although a few gyrs could be obtained in migration through the Netherlands and in Denmark to meet the European demand, most were acquired in the arctic and subarctic parts of their range. Trappers were sent at considerable cost, time, and difficulty, into remote areas to acquire them. Many of these expeditions were typically regulated and conducted under royal privilege (Allsen 2006).

European gyrfalcons were trapped in the Mountains of Norway and Sweden, and in Iceland. Annual falcon-catching expeditions were sent out to Iceland (Vaughan 1992), and at the time of the Vikings and the Medieval Warm Period, gyrs were obtained from Greenland. There were also falcon-trapping stations in Finmark and Troms in northern Norway that were rented out by the king (Vaughan 1992).

The birds acquired from Iceland and northern Scandinavia were cleared and tallied through the Danish court in Copenhagen (Dementiev 1960). According to Pertti Koskimes (2011) more than 6,200 gyrfalcons were shipped from Iceland to Copenhagen from 1664–1806, peaking every ten years. From 1731 to 1793 alone, 4,848 gyrfalcons shipped from Iceland to Denmark were tallied (Aegisson 2015). Olafur Nielsen and Gunnlaugur Petursson (1995) also tracked exports from Iceland to Denmark and similarly found a ten-year pattern of predominantly grey birds. [There was a much smaller export of white gyrs that were presumably migrants from Greenland but there was no cyclic pattern]. Georgiy Dementiev (1960) reported that from 1751 to 1764 as few as 103 and as many as 205 Icelandic gyrs were exported to Denmark per year. In 1754, one ship alone brought 148 gyrfalcons from Iceland to Denmark (Lloyd's Scandinavian Adventures, in Salvin and Brodrick 1855, cited in Shrubb 2013).

In Russia, gyrfalcons were under constant trapping pressure from the thirteenth to the eighteenth centuries (Dementiev 1960; Shergalin 2011). Much of this traffic in gyrfalcons was regulated by the aristocracy and consigned to local professional gyrfalcon trappers, known as Pomytchiki. Dementiev (1960) noted that from 1294 to 1304, the grand duke sent falconers to obtain gyrfalcons, primarily to preferred catching sites in the areas surrounding the White Sea and on the Murman coast. This included the Seven Islands, but also east along the Kanin Peninsula and the Pechora River and into the Ural Mountains, extending north along the southern part of the island of Novaya Zemlya. According to Allsen's (2006) research, Russia was considered a "central link in the transcontinental gyrfalcon network." One district east of the Yenisei Valley was so identified with the gyrfalcon industry that it was called, in Persian sources, the "Land of the Falconers."

By the sixteenth century, Russian gyrs were declared the property of the tsar. The harvest was regulated, and organized expeditions were sent out annually to meet the diplomatic requirements of Moscow

(Dementiev 1960). By the mid-sixteenth century, there were as many as 868 professional gyrfalcon trappers delivering gyrfalcons to Moscow, says Shergalin (2011). In the early seventeenth century, two ships were sent out each year, each to a different region in northern Russia, for gyrfalcon trapping.

Although this harvest in Russia dropped to fifty gyrfalcons per year at the beginning of the seventeenth century, it peaked again in the last part of that century with the reign of Alexei Mikhailovich Romanov, the tsar and grand duke of Russia from 1645-1676, himself an ardent falconer (Dementiev 1960). During his reign, there was a state monopoly on the acquisition of falcons. Falconers were sent north, as "a commandment of the Czar" to the well-known catching areas in the northwest part of Russia, but also further east. In 1652, the first organized expedition was sent into western Siberia. Here only gyrs were to be caught. This job was consigned to the indigenous Tatars who, for their effort, were freed of taxes. There was a quota on the grey gyrs, but not for the highly sought white birds, "as many as God gives shall be caught" (Dementiev 1960).

Gyrfalcons were so valued that one of the world's first nature preserves—the Seven Islands in Russia—was established during Mikhailovich's reign (1645-1676) specifically for the conservation of gyrfalcons (Ratcliffe 2005; Shergalin 2011). Here, gyrs were known to breed and overwinter (Dementiev and Gortchekovskaya 1945, in Potapov and Sale 2005).

Expeditions to obtain gyrfalcons were tightly regulated and elaborate (Dementiev 1960). Special legislation was enacted in the sixteenth century that defined the duties and behavior of trappers (Shergalin 2011). The falcon catchers were under strict instructions as to the catching, maintenance, and delivery of birds. They were prohibited to drink, to smoke tobacco, and to throw dice (gamble), so that "nothing untoward would happen to the Imperial birds." The gyrs were transported on special enclosed sleighs lined with felt and mats of braided linden or birch bark, or in special boxes lined with sheepskin and fastened to the sleigh, with 3-4 birds per sleigh. There were even special instructions on loading the boxes and to ensure a slow and careful journey. One cannot underestimate the value of these gyrfalcons for diplomatic purposes in these times of political instability.

The Chinese also went north in search of gyrfalcons. Thomas Allsen (2006) claimed, "People of the far north supplied prized raptors to the courts of China for nearly a thousand years." Marco Polo (1324), perhaps referring to the Bering Island within the commander group east of Kamchatka, describes one voyage that took forty days from China to the Great Sea Ocean, where gyrfalcons occur: "Upon traveling for forty days, as it is said, you reach the ocean...In an island lying off the coast, gerfalcons are found in such numbers that his majesty may be supplied with as many of them as he pleases. It must not be supposed that the gerfalcons sent from Europe for the use of the Tatars are conveyed to the court of the Great Khan. They go only to some of Tatar or other chiefs of the Levant, bordering on the countries of the Cumanians and Armenians. This island is situated so far to the north that the North Star appears to be behind you, and to have in part a southerly bearing."

The import of gyrfalcons into China began as early as 932 when thirty gyrfalcons were received by the Qitans (Liao Dynasty) from the Northern Pacific (from Wittfogal and Feng 1949, cited in Allsen 2006). The Qitans then initiated a tribute system that compelled the northern Jurchen tribes to acquire gyrs from the five Siberian Nations north of the Amur River (Allsen 2006). The exchange of gyrfalcons was eventually organized by way of an elaborate system of twenty-four relay stations called Falcon Stations, the most northerly one on the lower Amur River (Polo 1324).

THE ROLE OF THE VIKINGS

To satisfy the growing demand in Europe for gyrfalcons, expeditions were sent north into the Arctic, but this was a risky undertaking. Travel on the high seas and into the Arctic was precarious, time consuming, and expensive. Ice flows, the lack of provisioning, few ports, and limited knowledge were common hazards of travel. This was particularly so in Greenland, a key source of the prized white gyrfalcon. However, two events improved access to these distant lands and may have made possible the steady supply of white gyrfalcons to European markets—the Medieval warm period and the emerging domination of the Vikings in northern Europe.

The Medieval Warm Period enveloped the North Atlantic from about 800-1300 (Lamb 1966), as evident from ice cores and tree ring analysis (Loehle and McCullough 2008) as well as sedimentary records and moraine deposits (Araneda et al. 2007, cited in Salby 2012). Anecdotal reports indicate that this warming trend had a profound effect on living conditions at the time, and was a major influence on human settlement (Fagan 2008). Crop production was possible much further north and ocean travel into the Arctic was easier (Salby 2012).

Taking advantage of improvements in ocean travel were the Vikings, a warrior society in Scandinavia that persisted from 790 to 1066 (Clements 2005). Most commonly recognized as pirates (the word Viking is an early Scandinavian name meaning "pirate"), they initially made their presence known by raiding poorly guarded medieval monasteries across northern Europe. Eventually they became traders and occupied many areas across Europe. They also established colonies in Iceland and Greenland.

The timing was perfect for the Vikings. Not only was the climate warmer, there was a power vacuum across much of Europe after the collapse of the Roman Empire (Perry 2013). Trade was disrupted and many areas were poorly defended. The Vikings capitalized on this disorder, enabled by their famous shallow-draught long-ship with its sturdy keel that allowed it to be sailed or rowed.

It was well known that Greenland was the home to white gyrfalcons and other arctic-sourced luxury products. These included polar bear hides, bowhead whale baleen (used in women's girdles), walrus tusks and hides (the tusks were popular in the Islam states and the hides produced rope which was the strongest known at the time and of value for docking ships), narwhale tusks (known as "whale-tooth" and popular as decorative handles of knives and swords), and furs (which were of superior quality in the north) (Seaver 2010). Access to these special trading commodities was likely an important consideration in the establishment of outposts in Greenland by the Vikings in 986.

The settlements in Greenland grew and by 1012, the Vikings had a monopoly on the shipment of luxury goods from these northern frontiers to European ports (Spielvogel 2008). The Vikings also widened their reach into the Canadian Arctic, making contact with Thule and perhaps even the Dorset people who predated the Thule (Armstrong 2012), in all likelihood to secure luxury goods. It is speculated that the Vikings from Greenland obtained gyrfalcons from Baffin Island (Francis 2011), and that their search for luxury goods extended into Canada's western arctic (Busch 2015).

Viking domination ended suddenly. Perhaps more centralized authority in Europe and an improvement of coastal defenses, combined with the rise of Christianity, led to the demise of these Nordic adventurers. Armed pilgrimages blessed by the pope and orchestrated by the Teutonic Knights may have been particularly devastating to the Vikings (Lindholm and Licolle 2007). Pagans were punished or executed, and slavery, the cornerstone of Viking domination, was prohibited by the Christians. And so ended the Viking era, and presumably the reliable source of white gyrfalcons from Greenland.

THE YUKON CONNECTION

Cliff with poles leading up to a gyrfalcon nest, Ogilvie Mountains

The zealous pursuit of gyrfalcons even extended west into the Tutchone and Han country of what is today known as the Central Yukon in northwestern Canada. In a remote part of the Ogilvie Mountains, I happened to come across two gyrfalcon nest sites where there was a cluster of poles leading to the edge of the nests. The eyries had been used by gyrfalcons over many years, possibly centuries, as was evident by the substantial layer of excrement extending from the nest. The spruce poles propped against the wall of the cliff directly below the eyries were obviously dragged there to enable someone to enter the nest. Based on the heavy growth of the Elegant Sunburst Lichens (Xanthoria elegans) on these poles and surrounding rocks, they had been there a very long time.

Lichens are unusual plants. They are a symbiotic relationship between a specific alga or blue green alga and a specific fungus. These two life forms work together: the fungus obtains nourishment through decomposition and the alga manufactures food from sunlight through its photosynthetic ability. Lichens can persist without water for many years and have incredibly slow rates of growth. This makes them good estimators of historic and even geological events, and a clue as to the age of the poles I encountered leading into gyrfalcon nests.

The Elegant Sunburst Lichen, the one vividly displayed on the poles and the surrounding rocks, grows at a rate of about 0.5 mm/year for the first century, then slows down (McCarthy and Smith 1995). At this rate of growth, a lichen ring that is only five cm in radius is a hundred years old. Not only do they indicate long-term use of these nesting sites (because this species of lichen is associated with guano), they provide a clue as to the age of the human intrusion into these sites.

I was unable to determine how long ago, or by whom, the poles had been propped in front of the gyrfalcon eyries in the Central Yukon. Unfortunately, I did not collect materials that could be aged with radiocarbon dating. I spoke to a number of indigenous people in whose territories the poles were found. No one I spoke to had any stories or re-collections about any direct commercial interest in gyrfalcons. I assume that gyrfalcons were raided from these nests to supply the Eurasian lust for gyrfalcons. Were these the telltale markers of the Russian obsession with gyrfalcons during the late 1700s in the days of Russian trade along the coasts of Alaska?

Close up of poles and lichens

EUROPEAN TRADING CENTERS

In Europe, much of the trade in gyrfalcons took place in the well-known trading centers in Belgium and the Netherlands. The area of extensive heath, moor, and fen, now known as Flanders, along the coast of the Netherlands and Belgium, was well known as a passage route for birds migrating north and south. This huge wetland attracted many birds, including hawks, eagles, and falcons, and with them falconers. One small Dutch village, Valkenswaard, became synonymous with falconry, and well known as a desirable place to trap passage peregrines (Tom Cade, pers. com.). The town became a source of expertly crafted falconry accessories and a trading hub (Shrubb 2013). It was here that annual auctions were held for the purchase and trade in falcons and hawks. Falconers from the courts of feudal lords and kings throughout Europe, and likely beyond, would gather annually in Valkenswaard strictly for the commerce of falcons and hawks (Shrubb 2013). The gyrfalcon traffic was impressive. Between 1731 and 1793, 4,649 gyrfalcon transactions were reported from Valkenswaard (van de Wall 2004). Another great port city in the Flanders, Bruges in Belgium, also had links to the early falcon trade (Calvo et al. 2008). It too was a source of passage birds, but also had ties to Norway and therefore access to gyrfalcons.

A falconry school in Malbork, Poland, established by the Grand Master of the Order of Teutonic Knights in 1390, was another clearing-house for gyrfalcons (Shrubb 2013; Sielicki 2009). From there at least 1,800 trained falcons, including gyrfalcons, were sent as official gifts to kings, emperors, and the pope between 1533 and 1569.

WAS THE GYRFALCON HARVEST SUSTAINABLE?

The collection, possession, and trade of gyrfalcons during the falconry era from about 1200 to 1800 were tightly regulated in most countries where they were obtained. The king's appointments controlled much of these activities in most jurisdictions. In Iceland alone, from 1281 to the beginning of the nineteenth century, as many as 403 laws were instituted to control the possession and trade of gyrfalcons, says Dementiev (1960). Penalties for infractions were typically harsh, but were they enough to sustain wild gyrfalcons from the commercial exploitation?

Gyrfalcons are rare in nature and those that were caught were a long way from the markets. The Siberian birds that made their way to Persia were sometimes in transit for a hundred days by boat and caravan (Allsen 2006). Many died along the way. Taking into account years of ptarmigan scarcity and therefore low gyrfalcon production, and high rates of natural mortality during those ptarmigan-poor years, it may not have taken much to over-harvest the gyrfalcon population. So how did gyrs fare when they were at the height of fashion?

A number of records suggest that gyr populations suffered during the medieval period when the trade was flourishing. Dementiev (1960) suspected that, "as a result of being hunted for centuries, the number of gyrfalcons on Iceland has apparently decreased." Also, quotas set by the King of Denmark were reduced from a hundred to as low as thirty in 1785 (Aegisson 2015). Similarly, in Russia there were notes indicating that gyrs had declined. In one letter by Ivan the III concerning his efforts to supply white gyrfalcons from Russia (Dementiev 1960), he said, "I used to have good gyrfalcons, but they have almost died out." And in Greenland there were suspicions, albeit without much evidence, that gyrfalcon populations were suffering. Kirsten Seaver, in her book *The Frozen Echo* (1996), wrote,

Difficult to catch even in Greenland, gyrfalcons were worth a fortune by the time they reached Europe; the Duke of Burgundy is said to have ransomed his son from the Saracens as late as 1396 for twelve Greenland falcons. In 1276, just when Archbishop Jon was fine-tuning his laments, the Norwegian king sent the English king a princely gift of three white and eight grey gyrfalcons, a large number of ermine pelts, and a complete whale's head with all the baleen still attached. In the spring of 1315, Edward II of England sent a man to Norway to buy falcons and hawks – hardly a sign that these birds were a glut on the market. On the contrary, by 1337 gyrfalcons had grown in such short supply that Bishop Hakon of Bergen was obliged to write to King Magnus that he had not been able to obtain either white or grey falcons from an unnamed "Scottish Page," and the situation was no better three years later. When the Bergen Bishop wrote in November of 1340 to report to King Magnus about tax collections, he noted that Raimundo de Lamena, who was supposed to receive falcons in payment for apothecary goods, had not been able to get more than two or three birds from the royal palace in Bergen. These incidents strongly suggest that both white and grey gyrfalcons which had been a prized Greenland export since the beginning of settlement, were valuable as ever in the second quarter of the fourteenth century. If anything, there was a shortage of supply (Kirsten Seaver, the Frozen Echo, Stanford University Press, Stanford, 1996: 82).

Many factors probably influenced the success of falcon catchers to secure gyrfalcons: low gyrfalcon densities, natural cycles, travel conditions, the number of falcon catchers, the difficulty of capture, the losses of birds in transit, etc. Or perhaps with the diminishing role of the Vikings after 1100 and the cooling of the climate after 1300, the steady stream of gyrfalcons was severely interrupted. Are the harvest records and anecdotal observations during the Middle Ages an indication of over-harvest, or do they simply reflect that gyrfalcons were never very abundant and perhaps difficult to gain access to and catch, more so in some years than others? Or does the trend simply reflect a declining interest in falconry?

What is remarkable is the apparent resilience of gyrfalcon populations in the face of sustained periods of heavy trapping. In Iceland in the seventeenth and eighteenth century, when the trapping of gyrfalcons was carried out under royal patents issued by the Danish Kings, 100 to 200 gyrfalcons were trapped and sent to Denmark annually (Nielsen and Petursson 1995). Based on thirty years of research in Northeast Iceland (Nielsen 2011) and population projections for the entire island (Potapov and Sale 2005), an average of 278 gyrfalcons would be produced in Iceland per year. If just 80% of the gyrs tallied under royal patent were from Iceland (assuming some were migrants, probably from Greenland), it represents 29 to 58% of the average annual production of gyrfalcons. As most of the trapping of gyrfalcons occurred early in the breeding season and was directed toward adults and first year survivors (no eyasses), the impact on the breeding population of gyrfalcons would be expected to be even greater (Tom Cade pers. comm.). Yet there is no evidence that this level of harvest had any appreciable effect on the breeding population. Tom Cade (1968) remarked that the ability of the gyrfalcon population to buffer such exploitation is likely because of the periodic production of a large reservoir of non-breeders—the floaters.

Falconers, artwork by Vadim Gorbatov, depicting father and son with gyrfalcon and merlin in the 17th century

THE ERA OF FALCONRY, IN RETROSPECT

The sport of falconry had a remarkable presence in Europe and Asia. It gained prominence at a time when the feudal system grabbed hold of Europe, and when the rituals of war were glorified. Falconry, "considered among the finest of earthly pursuits" (International Association of Falconry and Conservation of Birds of Prey 2005), was appealing because, to many, it is a very thrilling and entertaining spectacle. As well, because the sport took time, money, and extensive training, it was a display of conspicuous indulgence.

Today most every country in Europe and Asia can bear witness to its tradition of falconry. Falconry was the height of fashion for over a thousand years, its practice contributing to language, legends, artifacts, literature, and art. Even the development of specific breeds of dogs and horses can be ascribed to falconry. Many of the English words we use today had their genesis in the jargon surrounding falconry, e.g., hoodwink, haggard, bated-breath, fed-up, chaperon, callow, booze, codger, gorge, and rouse (Pepper 2012). And so gyrfalcons, "the one who stays all winter" in the land of darkness, scarcely venturing south of the sixtieth parallel and largely unknown by much of the human population today, were, during the Middle Ages, renowned around much of the world as an icon of prestige and power. Allsen (2006) suggested that the acquisition of gyrfalcons brought northern peoples into tribute relations with imperial powers, "In such ways [that] the ethnic histories and subsistence systems of faraway, small-scale societies became entwined with diplomatic practices of great powers and the transcontinental cultural exchanges of great civilizations."

CHAPTER TEN

FROM VENERATION TO OPPRESSION TO GLORIFICATION

FALCONRY EVENTUALLY LOST PRESTIGE ACROSS EUROPE AND much of Asia. Its demise, together with changing attitudes towards birds of prey, was the outcome of four significant social changes: the disintegration of feudalism, the expanding "middle class," the industrialization of farming, and the introduction and widespread use of the shotgun.

The warmer climate in northern Europe that had helped sustain feudalism did not last. Enter the Little Ice Age after 1300. Temperatures dropped below previous norms, then became increasingly cold from 1560-1850 (Fagan 2000). Shorter growing seasons with harsh winters resulted in widespread crop failures, abandonment of farms, substantially reduced grain harvests, and poor seed production (Lamb 1995).

The common-people began to suffer. As well as being hungry and in poor health, they were heavily taxed by both the state and the Roman Catholic Church (Alink and van Kommer 2011). To add insult to injury, the nobility and the clergy were typically exempt from taxes, and the noble class continued to display their affluent indulgences. The combination of food scarcity, excessive taxes, widespread poverty, and the obvious disparity in wealth caused civil unrest that eventually exploded with the French Revolution in 1789 to overthrow the aristocracy. This put an end to the absolute monarchy with its extravagant feudal privileges for the nobility and clergy. Similar revolts, albeit less violent, spread across Europe. Falconry, the badge of affluence, became a social disgrace.

The general disintegration of the feudal system and the redistribution of wealth and land, together with the industrialization of farming, had two effects. Much of the land where falcons once flew as a spectacle to the nobility was now cleared and cultivated. The peasants and the gentry of the feudal system became part of a growing middle class, and more farms sprung up. Birds of prey were no longer valued as entertainment, but viewed as a threat to domestic fowl, and were shot or trapped.

The increasingly widespread use of the shotgun (fowling piece) for game hunting in the eighteenth and nineteenth centuries further threatened birds of prey. Not only did it make killing raptors easy, the shotgun led to the hunting of grouse, waterfowl, and rabbits as food and sport. The upper classes replaced falcons with shotguns. Sport hunting became the pastime of the rich and idle. Gyrfalcons and other birds of prey represented a threat to these activities. In North America, the killing of eagles, hawks, and falcons was encouraged, and in some areas, it was promoted through bounties (Carrie 2013). Many birds of prey were simply called "chicken hawks" and few people could even identify the various species that they shot. All were classified as vermin. Gyrs were not spared. In Iceland, a law in 1885 allowed gyrs to be killed at any time and with no limitations, (Aegisson 2015). Andreas Hagerup (1891) describes a hunt in Greenland: "In the winter I used to take my pigeons out every day, so that when a falcon came in sight I might induce them to come within shooting range."

There was also conflict with pigeon racing. Pigeon racing became popular in Flanders in the mid-nineteenth century, and spread across Europe and into North America. It was lucrative and led to intensive breeding for superior racing birds. Some falcons were good at intercepting racing pigeons despite selective breeding for exceptional speed. Falcons threatened the races and the investments toward superior pigeons and consequently were persecuted by pigeon fanciers (Enderson 2005).

In some countries, egg collectors and museum collectors also took a toll on raptor populations. The gyrfalcon was given no special treatment. Gyrs were prized specimens valued for their rarity. Along the Greenland coast in 1928, at least 250 gyrfalcons were shot (Salomonsen 1950) during a period of collection that continued for many more years. Universities, museums, and private collectors all participated in the collection of gyrfalcon skins. In 1929, the University of Michigan alone acquired eighty-one gyrs from Greenland (Vaughan 1992). The museum in Copenhagen (Denmark) holds about eight hundred specimens of gyrfalcons, most from Greenland (Potapov and Sale 2005). Gyrfalcon eggs were also collected. Vaughan (1992) commented that the removal of eggs and birds from Iceland in the early 1900s may have seriously reduced the gyrfalcon population, and led to their full protection by the Danish authorities in 1913.

But public attitudes towards raptors eventually changed. The impetus to save birds of prey in North America first gained strength with the tireless campaign of Rosalie Edge to stop the shooting of hawks migrating along the Appalachian ridge in East Central Pennsylvania. Her efforts led to the establishment of the Hawk Mountain Sanctuary in 1934—the first refuge for birds of prey in North America (Furmansky 2009).

Aldo Leopold (1949) and Rachael Carson (1962) also tried to swing us back to nature and convince us that the land and its inhabitants, including predators, had intrinsic value. Through his writings, Leopold introduced settlers and immigrants to a type of land ethic that indigenous societies were built upon. He pleaded with his audience to "think like a mountain"—that the parts of our environment are connected and inter-dependent. In North America, he is considered the father of wildlife management. Carson followed in these teachings, questioning the widespread use of pesticides and the disregard for the ecosystem. Soon after the environmental movement emerged, wildlife management agencies sprung up and new policies developed in North America. A similar land ethic emerged in Europe. The killing of gyrfalcons was again prohibited over much of their range.

Today the gyrfalcon is again revered in North America, not just for falconry or as a cultural icon, but as a fragile and uncommon apical predator. Ecologists often refer to the gyr as a keystone species, giving them the same notoriety that they reserve for polar bears and grizzly bears. The fact that gyrs are rare, beautiful, and have a long fascinating history of association with humanity, further elevates their status.

THE REBIRTH OF FALCONRY AND THE ICONIC GYRFALCON

The widespread growth of farming following the Industrial Revolution did not play out well for most raptors. Loss of habitat and direct persecution, including bounties, were not the only agents in their demise. Cultivation was accompanied by the excessive use of pesticides, some of which persisted in the environment and accumulated at the top of the food chain.

These were the silent killers. DDT (dichlorodiphenyltrichloroethane) in particular had a devastating effect on some avian predators, notably peregrines, bald eagles, and ospreys, by altering their calcium metabolism. This caused eggshell thinning to a point where they could not withstand the pressure of the incubating parent (Ratcliffe 1967; Hickey and Anderson 1968). All eyes were on the peregrine falcon. Populations in the eastern US and southeastern Canada were exterminated, and northern and western populations were quickly disappearing.

Falconers and biologists teamed up to propose a solution. They suggested that if peregrine falcons could be bred in captivity, this effort could produce a source of nestlings that could be introduced into the wild as a way to compensate for egg losses, as well as provide a source of birds for falconers, thereby reducing the motivation to take falcons from the wild (Cade and Burnham 2003). The idea gained traction with the help of the Raptor Research Foundation, which was founded in 1966 in large part to promote research dedicated to the management and recovery of raptors. As a result, many successful breeding projects were initiated across the US and then into Europe and eventually into the Middle East (Cade and Berry, in prep.). Even gyrfalcons played a role in the peregrine recovery. In the northern Yukon, peregrine chicks were placed into active gyrfalcon nests and successfully fledged (D. Mossop, pers. comm.).

Tom Cade, a key player in the recovery of peregrines, with his gyrfalcon

Peregrine falcons became the poster child for conservation. Earnest efforts to save peregrines, including the establishment of The Peregrine Fund in 1970 by Tom Cade, were very successful. Peregrine populations rebounded, they returned to their old haunts, and in 1999, they were taken off the endangered species list in the US and later in Canada (Carrie 2013). In the course of their return to the wild, peregrines received much publicity, raising awareness for birds of prey, including the iconic gyrfalcon.

At the same time that captive bred peregrines were being introduced into the wild, interest in falconry rebounded. Perhaps our inherent desire to reconnect with nature, combined with more free time, stimulated the age-old tradition of falconry—the quintessential bond between humanity and the natural environment. Aldo Leopold (1949), who reminded us, "Wilderness gives definition and meaning to the human enterprise," viewed falconry as the most glamorous and perfect hobby.

With the rising interest in flying hawks and conservation in general, many countries legalized falconry after decades of prohibition. In 2010, the United Nations Educational, Scientific, and Cultural Organization (UNESCO), through its Convention for the Safeguarding of the Intangible Cultural Heritage, listed falconry as a living cultural heritage.

Although the driving force was to restore depleted populations of peregrines and other birds of prey, captive breeding has continued to grow to meet the increasing demands of falconry, perhaps stimulated by the UNESCO listing and advancements in breeding techniques. Tom Cade and Robert Berry (in prep.) have compiled a number of estimates that underline this growth in falconry and captive breeding. They claim there are probably 50,000 to 70,000 falconers and more than 6,500 breeding facilities worldwide that produce over 35,000 birds of prey a year to satisfy at least half of the requirements for falconry. Gyrfalcons and their hybrids alone are now produced in the thousands.

Falconer attempting to trap a gyrfalcon

The challenge today, suggests Cade and Berry, is not providing enough falcons to meet the demand for falconry, but rather conserving enough land to practice the original sport. Human growth and the continued fragmentation of wilderness mean fewer places to fly birds in the traditional ways. The original practice has largely been replaced with other forms of recreation that involve birds of prey. Many birds are now flown to compete in races, speed trials, pigeon derbies, and sky trials (competitions that measure the height gained by the falcon in advance of the attack by using lures attached to helium balloons or drones) says Cade and Berry, adding that in some countries, natural quarry has even been replaced with simulated robotic prey flown to entice raptors.

However, keeping alive the original practice of falconry, which has gained recognition as a living cultural heritage, is not the only concern. Despite our success in meeting the demand for falconry and providing a source of birds for relocation into the wild, there is widespread concern today for the conservation of wild gyrfalcons (and other birds of prey) from the perils of climate change.

CHAPTER ELEVEN

IS THERE A FUTURE FOR GYRFALCONS?

GYRFALCONS HAVE GENERALLY BEEN ISOLATED FROM DEVELOPMENT, contaminants, and human-caused ecological changes. They live in remote regions and for the most part eat locally. For the last fifty or so years, gyrs have also been protected through international treaties and national laws across most of their range.

However, the future for wild gyrfalcons is uncertain. Climate change, one of a number of symptoms of our mismanagement of the planet, is creating uncertainty about ecological balance. We have been warned that there will be major changes in temperature and precipitation, accompanied by melting permafrost, advancing tree line, rising ocean levels, and changing water tables. We have been told to anticipate severe drought in some areas and an increase in snowfall in other areas, and we have been warned that times will be unstable—more catastrophic weather events and irregular weather patterns. Indeed, we are now experiencing most of these predictions.

But it is more than dramatic ecological changes that threatens gyrfalcons. The human population is over seven billion and continues to grow, putting tremendous pressure on the natural resources that we require. The pressure is compounded by our need for more and more commodities, and our expectation that the economy must grow. To satisfy our insatiable demands, the environment is typically sacrificed. The North, with its untapped resources and improved access because of changing ice patterns, is attracting much attention. Gyrfalcons may soon be at the mercy of industrial development, as well as significant ecological changes, including exposure to new diseases.

The gyrfalcon, at least in the Yukon, has simple needs: ptarmigan and nesting cliffs in open landscapes, mediated by some exacting timing requirements (the courtship of ptarmigan and possibly the dispersal of ground squirrels), and interactions with golden eagles. How will climate change alter these dynamics?

The trends are not reassuring. Tom Cade compiled a list of those areas where gyrfalcons are suspected of being in decline or have disappeared. He includes Labrador in North America, as well as Southern Kamchatka, the Komandorskye Islands, the middle latitudes of the Ural Mountains, and South and West

Greenland (Cade 2011). Ornithologist Dave Mossop (2011) suspects that gyrfalcons in the Yukon are also declining.

However, there are some exceptions. Gyrfalcons appear to be on the increase on the Southern Yamal Peninsula in Northwest Siberia, and in some of the wetlands east of the Ural's (Mechnikova et al. 2011). In northern Fennoscania, gyrfalcon populations appear stable when compared with data collected 150 years earlier (Koskimes 2011).

Ptarmigan may be the direct victims of climate change and driving the decline or redistribution of gyrfalcons. Rock ptarmigan have been in decline in Fennoscandia since the 1990s (Koskimies 2011) and similar trends are evident in Alaska (Martin and Wilson 2011). The Alaskan researchers suggested that ptarmigan suffer from a combination of unpredictable events, including freeze-thaw conditions that appear to reduce their overwinter survival (possibly by coating the willow buds with ice), and delays in snowmelt that postpone breeding and lower reproductive success. A team of researchers in northern Sweden also forecasted the decline of rock ptarmigan as due in part to the loss of permanent snow fields (Pedersen et al. 2014), and a Russian biologist suggested that unpredictable weather events in Northeastern Siberia may be devastating to ptarmigan (Isaev 2011). Changes in ptarmigan numbers will undoubtedly affect gyrfalcons, but are there other factors that will come to bear?

What about the effect of climate change on the dynamics of the gyrfalcon population cycle itself? Population modeling of snowy owls and collared lemmings, through twenty-year time series analysis of population density, predicted that climate change would extend the length of the lemming cycle and reduce the maximum population density of lemmings (Gilg et al. 2009). The authors surmised that this change would have an adverse impact on the abundance of snowy owls, the populations of which are adapted to capitalize on the pulse of cyclic lemming populations. Gilg and his team (2009) speculated that this impact might be an early sign of severe impact to the predator-prey communities in the Arctic from climate change. Indeed, the population of collared lemmings in Greenland today is at low density and appears to be non-cyclic, coinciding with a 98% decline in snowy owl fledgling production, prompting concerns that the lemming cycle will collapse and snowy owls will disappear (Schmidt et al. 2012). In North America, there are also mounting concerns about the disappearance of snowy owls. The recent Partners in Flight Report, produced by a team of authors from academic, government, and conservation institutions in Canada and the U.S., indicated a 74% drop in the number of snowy owls from 1970-2014 (Andrew-Gee 2016). Is this decline also the consequence of a collapsing lemming cycle induced by climate change?

Recent findings have also found disturbing changes in the dynamics of the ptarmigan cycle, including disruptions in its regularity, wider amplitudes, a dampening of peaks, and in one case, the disappearance of the cycle itself (Newton 2011). In some areas in the Yukon, the ptarmigan cycle appears to be faltering and the peaks are disappearing, depriving gyrfalcons of that periodic surge in food supply (Mossop 2011). Although climate change is thought to be the cause, the agents of these adjustments in the ptarmigan

Glaciers re-ceding in the Selwyn Mountains, Yukon

cycle, and its corresponding effects on gyrfalcons, are unknown. Are these changes the outcome of random, unpredictable weather events that are modulating the cycle? If catastrophic weather events affect ptarmigan population dynamics and if these weather events are occurring more often, can this account for changes in the ptarmigan cycle, including the dampening of its peaks?

Perhaps more worrisome for gyrfalcons is the accentuation of the trough in the ptarmigan cycle. If ptarmigan populations fall below a certain threshold, gyrs may be unable to remain in the area, as appeared to be the case in one region in Russia (Morozov 2011). Even small changes in the depth or persistence of these ptarmigan troughs, may affect gyrfalcons if they are unable to contribute enough youngsters to compensate for adult losses.

In some parts of their range, climate change may also be degrading gyrfalcon nest sites and increasing competition for food. In the southern part of their breeding range in Russia, where 80% of nesting pairs are found in trees, freeze-thaw patterns are causing ice build-up in large stick nests, forcing gyrfalcons to find alternative nests, most often in less favorable situations (Mechnikova et al. 2011).

Glaciers receding in the Selwyn Mountains, Yukon

There is also a concern that climate change may be favoring golden eagles. Researchers in Northern Fennoscandia found that the expansion of birch forests was associated with an increase in nesting golden eagles, and they predicted that with greater numbers of eagles, gyrfalcons will suffer (Johansen and Ostlyngen 2011).

Gyrfalcons may also fare poorly with the expansion of the range of peregrine falcons. A dramatic increase in the number of peregrine falcons in Western Greenland in the last fifty years may be squeezing gyrfalcons out due to competition for food and possibly nesting cliffs (Burnham and Burnham 2011).

Perhaps more threatening is the prognosis for more parasites and diseases. Northern environments are relatively simple (fewer species and fewer ecological networks). Those species that live in the north year-round, including gyrfalcons, have low immunity to many of the pathogens found further south. Climate change may be making possible the northerly range expansion of many species of insects, bacteria, and viruses. The gyrfalcon will be exposed to many new parasites and pathogens, the rapid invasion of which could be catastrophic.

Indeed, the devastating effects of parasites and diseases on northern ecosystems are increasingly obvious. The western pine beetle has ravaged pine forests through much of Western Canada, and the spruce bark beetle and the spruce budworm, both benefitting from warmer winters, are now major forest pests in Alaska and the Yukon.

There is another potential impact of climate change that is becoming apparent in Central Asia and Siberia, according to Tom Cade. In some areas, the Siberian Steppe is gradually moving northward, encroaching into the Arctic tundra. This may result in a rearrangement of the distributions of prey, which may bring gyrfalcons and saker falcons together. It is well known that these two species of falcons can interbreed and produce fully fertile offspring. If habitat changes favor the coexistence of the two species, it may create a hybrid zone where gyrfalcons will be replaced with a population of gyrfalcon-saker hybrids (T. Cade, pers. comm.).

Solutions to the problems associated with climate change are not obvious. Nations may accept that the climate is changing, but they rarely agree on how to allocate responsibility for addressing the problem. Most governments seem unwilling to tinker with an economic system that promotes the largely unregulated exploitation of resources, and most individuals participating in the capitalist economy appear unprepared to consume less. As Aldo Leopold remarked back in 1949, "Society is now like a hypochondriac, so obsessed with its economic health as to have lost its capacity to be healthy." Quality of life has been supplanted by our desire for wealth, or "standard of living," and with the global population rapidly increasing, more people are laying claim to its riches.

There are other risks to gyrfalcons. Poaching continues to be a serious concern in some parts of Russia (Potapov 2011b) and there are concerns about the increasing prevalence of environmental pollutants reaching the north (Braune 2011). But these issues are dwarfed when we contemplate a situation of major ecological imbalance, where changes come too quickly to afford the gyrfalcon or ptarmigan the chance to adapt.

If we are to find a balanced way to live on the planet, I think we must return to a more indigenous holistic worldview, one based on respect for the environment and mindful of what we are leaving our grandchildren. In this indigenous worldview, land is much more than a collection of commodities to serve our insatiable "needs," it has intrinsic, sacred value. Aldo Leopold (1949) espoused a similar doctrine of respect and interconnection with nature, pleading with us to adopt what he called a "land ethic" or "ecological conscience." Leopold argued that without a land ethic our relationship with land "is strictly economic, entailing privileges but not obligations." The consequences of such a narrowly conceived economic model are evident today.

I think the late Steven Jay Gould, the brilliant paleontologist, evolutionary biologist, and writer, in the prologue of his collection of essays published in 1991 under the title *Bully for Brontosaurus*, had the right insight into the future of our species, and by extension, the gyrfalcon:

Our planet is not fragile at its own time scale, and we, pitiful latecomers in the last microsecond of our planetary year, are stewards of nothing in the long run. Yet no political movement is more vital and timely than modern environmentalism—because we must save ourselves (and our neighbor species) from our own immediate folly. We hear so much talk about an environmental ethic. Many proposals embody the abstract majesty of a Kantian categorical imperative. Yet I think that we need something far more grubby and practical. We need a version of the most useful and ancient moral principle of all—the precept developed in one form or another by nearly every culture because it acts, in its legitimate appeal to self-interest, as a doctrine of stability based upon mutual respect. No one has ever improved upon the golden rule. If we execute such a compact with our planet, pledging to cherish the earth as we would wish to be treated ourselves, she may relent and allow us to muddle through. Such a limited goal may strike some readers as cynical or blinkered. But remember that, to an evolutionary biologist, persistence is the ultimate reward. And human brainpower, for reasons quite unrelated to evolutionary origin, has the damnedest capacity to discover the most fascinating things, and think the most peculiar thoughts. So why not keep this interesting experiment around, at least for another planetary second or two?

LOOKING AHEAD

THE YEAR 2050: IT WAS FALL AND the sun was waning in the sky. However, the seasons did not come and go as they did in the past. In some years, summer would persist well into the period when the daylight hours were decreasing, but in other years winter grabbed hold early and forcefully. Winters too were not the same. Often a warm air would temporarily envelope the land, melting the surface of the snow, which would freeze and form a crust. Winter rains could also be expected now, the effects of which would leave willow buds and stems encased in ice. The brutal cold was no longer the norm but rather came intermittently between the warming spells. Snow too was typically deeper and heavier, and often layered with ice that attested to the periodic freeze-thaw conditions. But then in some years, normality returned and winters were cold, the snow was soft and buoyant, and fall and spring arrived on schedule.

The land reflected these climatic changes. The willow shrubs were taller and they spread further up the mountain, replacing many of the prostrate shrubs. The many ponds that characterized the tundra were disappearing, no longer suspended by the permafrost that was melting. The stunted spruce were also invading the highland, in clumps and fingers that reached beyond the boreal forest. Many of the creek bottoms were now scoured, a result of more rapid discharge of water during the spring runoff.

The land continued to provide for the gyrfalcon, but not as it did in the past. The ptarmigan were less common, and their cycles were less dramatic and less predictable. They suffered from the periodic icing of the willow buds and the late snows that mocked their plumage change making them more conspicuous. The wetlands, the heart of the ecosystem, were less common and less productive. Fewer water bodies meant fewer water birds.

But the magic of spring still aroused this old hunter, and the summer was still dynamic. The gyrfalcon ventured farther to feed his family, adjusting to the changes, all the while avoiding golden eagles. Eagles were becoming more common with the proliferation of arctic ground squirrels whom were benefiting from more suitable denning sites made possible with the melting permafrost. The wetlands, although muted and more widely scattered, continued to provide the pulse of life and excitement, and the flowers continued to bring color to the uplands.

Despite these environmental changes, the gyr did not lose his ancestral urge to hunt and reproduce. As the climate changed, he could count on two things that were unaltered: his remarkable ability to hunt and his vast reservoir of experience. These capabilities would allow him to adapt to this rapidly changing world.

For now, he was content to enjoy this moment, perched here looking over the ridges now splashed with yellow, red, tangerine, and burgundy, and absorbing the last little bit of heat from the autumn sun. Winter would be here soon—silent and pure but dark and desolate—and it would again test his capabilities. *The challenges of winter would only heighten his anticipation for the spring, when nature would resurrect itself, and hopefully, he would survive to endow the world with more gyrfalcons.*

PHOTO CREDITS

After the Table of Contents, Pages vi; xii; 2; 20; 56; 58; 62; 73; 140 (top), courtesy of Bryce Robinson

Pages v; x; xi; 89; 139, courtesy of Jannik Schou

Pages viii; 5; 8; 9; 10; 12; 15 (top and bottom); 16; 18 (top and bottom); 19; 21; 22-23; 25; 28; 31 (top and bottom); 34 (top and bottom); 35; 36; 37; 38; 39; 41 (top and bottom); 42 (top and bottom); 43 (top, bottom left, bottom right); 44; 45; 47; 49; 50; 52; 53 (top and bottom); 54; 59; 68; 74; 75; 78; 79; 83; 85; 87; 124; 131 (left and right); 133; 137; 140 (bottom), Norman Barichello

Page 3, courtesy of David Anderson

Page 26, remote camera image, courtesy of Michael Henderson and The Peregrine Fund

Pages 29; 98; 106; 110; 112-113; 116-117; 127, courtesy of Vadim Gorbatov

Pages 33 and 136, courtesy of Joshua Barichello

Pages 64, 69, 70, 71, 88, remote camera images, courtesy of Bryce Robinson and The Peregrine Fund

Pages 92-93, courtesy of Abraham Anghik Ruben, and Rocco Paneese with the Kipling Gallery

Page 94 (top left and right), courtesy of Alexander Stubbing, Edward Atkinson and the Nunavut Heritage Centre

Page 95, courtesy Pixabee

Page 101, courtesy of the Victoria and Albert Museum

Page 123, courtesy of Dave Mossop

Page 132, courtesy Kate Davis

Page 161, courtesy Barbara Gale

REFERENCES

Aegisson, S. 2015. Icelandic trade with gyrfalcons, from medieval times to the modern era. Self-published by Sigurdur Aegisson. Siglufjorout, Iceland.

Alikuzai, H.W. 2013. A concise history of Afghanistan in 25 volumes. Trafford Publishing, on-line publishers.

Alink, M. and V. van Kommer. 2011. Handbook on Tax Administration. IBFD, Amsterdam, The Netherlands.

Allsen, T.T. 2006. Falconry and the exchange networks of medieval Eurasia, pages 135-155 in Pre-Modern Russia and its world: Essays in honor of Thomas S. Noonan. Ryerson, K.L., T.G. Stavrou and J.D. Tracy (eds). Harrassowicz Verlag, Germany.

Allsen, T.T. 2011. The Royal Hunt in Eurasian History. University of Pennsylvania, Philadelphia, Pennsylvania.

Alunik, I, E.E. Kolausok and D. Morrison. 2005. Across Time and Tundra – The Inuvialuit of the Western Arctic. Raincoast Books, Vancouver, and Canadian Museum of Civilization.

Andreev, A.V. 1990. Winter adaptation in willow ptarmigan. Arctic 44(2): 106-114.

Andrew-Gee, E. 2016. Bird populations in steep decline in North America, study finds. The Globe and Mail, September 14, 2016.

Araneda, A., F. Torrejon, M. Aguayo, L. Torres, F. Crueces, M. Cisternas, and R. Urritia. 2007. Historical records of San Juan glacier advances: another clue to Little Ice Age timing in Southern Chile. The Holocene 17: 987-998. Cited in Salby 2012.

Armstrong, J. 2012. Vikings in Canada? A researcher says she's found evidence that Norse sailors may have settled in Canada's Arctic. Others aren't so sure. Maclean's Magazine, November 20, 2012, page 26-27

Asbridge, T.S. 2004. The First Crusade: a new history. Oxford University Press, Oxford, UK.

Auger, E.E. 2005. The way of Inuit Art – Aesthetics and History in and Beyond the Arctic. McFarland and Company Inc., Jefferson, North Carolina.

Barichello, N. 1983. Selection of nest sites by Gyrfalcons. M.S. thesis, University of British Columbia, Vancouver, British Columbia, Canada.

Barichello, N. 2011. Gyrfalcon Courtship – a Mechanism to Adjust Reproductive Effort to the Availability of Ptarmigan. Pages 339-354, Volume I, In R.T. Watson, T.J. Cade, M. Fuller, W.G. Hunt and E. Potapov (Eds.), Gyrfalcons and Ptarmigan in a Changing World, The Peregrine Fund, Boise State University, Idaho, USA, 1-3 February 2011.

Barichello, N. and D. Mossop. 2011. The Overwhelming Influence of Ptarmigan Abundance on Gyrfalcon Reproductive Success. Pages 307-322, Volume I, In R.T. Watson, T.J. Cade, M. Fuller, W.G. Hunt and E. Potapov (Eds.), Gyrfalcons and Ptarmigan in a Changing World, The Peregrine Fund, Boise State University, Idaho, USA, 1-3 February 2011.

Bartoslewiez, L. 2012. Show me your hawk, I'll tell you who you are. In D.C.M. Raemaekers-E. Esser, R.C.G.M. Lauwerier, J.T. Zeiler, eds: A Bouquet of Archaeozoological Studies, Essays in Honour of Wietske Prummel. Groningen, Bakhuis and University of Groningen Library.

Beebe, Frank. 1976. Hawks, Falcons and Falconry. Hancock House Publishers Ltd. Saanichton B.C. Canada 320 pages.

Belon, Pierre. 1555. L'Histoire de la nature des oyseaux, avec leurs descriptions, and naifs partaicts retirez du naturel: escrite en sent livres. Cited in Dementiev (1960).

Beni, A. 2016. Turul, the Mystical Hungarian Mythological Bird. Daily News Hungary, April 4, 2016.

Bente, P.J. 1981. Nesting behavior and hunting activity of the Gyrfalcon, Falco rusticolus, in South Central Alaska. M.Sc. Thesis, U. of Alaska, Fairbanks, Alaska USA. 103pp.

Berezkin, Y.E. 2015. Siberian folklores and the Na-Dene origins. Archaeology, Ethnology and Anthropology of Eurasia, 43(1).

Bergerud, A.T. 1988. Mating systems in grouse. In A.T. Bergerud and M.W. Graton (eds.), Adaptive strategies of population ecology of northern grouse. U. of Minnesota Press.

Bergerud, A.T. and M.W. Gratson. 1988. Adaptive Strategies and Population Ecology of Northern Grouse: Volume 1. Population Studies. University of Minnesota Press. 444pp.

Berners, J. 1486. The Boke of St. Albans. Reprinted by BiblioBazaar, August 2015. 218pp. Cited in Bartoslewiez 2012.

Bo, O. 1962. Falcon Catching in Norway, with the Emphasis on the Post-Reformation Period. Universitetslaget, Oslo. Cited in Potapov and Sale 2005.

Bodio, S. 2009. Deep Roots. In Cutchins, D. and E.A. Elliason (eds.). 2009. Wild Games – Hunting and Fishing Traditions in North America. University of Tennessee Press, Knoxville, Tennessee, USA.

Booms, T.L., and M.R. Fuller. 2003a. Gyrfalcon diet in central and West Greenland during the nesting period. Condor 105:528-537.

Booms, T.L. and M.R. Fuller. 2003b. Gyrfalcon feeding behaviour during the nestling period in central west Greenland. Arctic 56: 341-348.

Braune, B.M. 2011. Chemical Contaminants in the Arctic Environment – Are They a Concern for Wildlife. Pages 133-145, Volume I, in R.T. Watson, T.J. Cade, M. Fuller, W.G. Hunt and E. Potapov (Eds.), Gyrfalcons and Ptarmigan in a Changing World, The Peregrine Fund, Boise State University, Idaho, USA, 1-3 February 2011.

Burnham, K.K. and W.A. Burnham. 2011. Ecology and Biology of Gyrfalcons in Greenland. Pages 1-20, Volume II, in R.T. Watson, T.J. Cade, M. Fuller, W.G. Hunt and E. Potapov (Eds.), Gyrfalcons and Ptarmigan in a Changing World, The Peregrine Fund, Boise State University, Idaho, USA, 1-3 February 2011.

Burnham, K.K. and I. Newton. 2011. Seasonal movements of gyrfalcons Include Extensive Periods at Sea. Pages 49-70, Volume II, In R.T. Watson, T.J. Cade, M. Fuller, W.G. Hunt and E. Potapov (Eds.), Gyrfalcons and Ptarmigan in a Changing World, The Peregrine Fund, Boise State University, Idaho, USA, 1-3 February 2011.

Burnham, K.K., W.A. Burnham, and I. Newton. 2009. Gyrfalcon Falco rusticolus post-glacial colonization and extreme long-term use of nest- sites in Greenland. Ibis 151:514-522.

Busch, L. 2015. The Historian. Up Here Magazine. July 2015. Pages 1-14.

Cade, T.J. 1960. Ecology of the Peregrine and Gyrfalcon populations in Alaska. U. of California Press. Berkley and Los Angeles California. 267pp.

Cade, T.J. 1968. The Gyrfalcon and falconry. The Living Bird 7:237-240.

Cade, T.J. 1982. The Falcons of the World. Cornell University Press, Ithaca, New York, USA.

Cade, T.J. 1953. Behavior of a young Gyrfalcon. Wilson Bulletin 6: 26-31.

Cade, T.J. 1951. Carnivorous Ground Squirrels on St. Lawrence Island, Alaska. Journal of Mammalogy, Volume 32 (3): 358–360.

Cade, T.J., P, Koskimies, and O. Nielsen. 1998. Falco rusticolus Gyrfalcon. Birds of the Western Palearctic Update 2: 1-25.

Cade, T.J. 2011. Biological Traits of the Gyrfalcon in Relation to Climate Change. Pages 33-44, in Volume I, In R.T. Watson, T.J. Cade, M. Fuller, W.G. H`unt and E. Potapov (Eds.), Gyrfalcons and Ptarmigan in a Changing World, The Peregrine Fund, Boise State University, Idaho, USA, 1-3 February 2011.

Cade, T.J. and R.B. Berry. In prep. The influence of propagating birds of prey on falconry and raptor conservation.

Calvo, E., M. Comes, R. Puig, and M. Rius. 2008. A Shared Legacy: Islamic Science East and West. University of Barcelona.

Campbell, B. and E. Lack. 1985. A Dictionary of Birds. T and A.D. Poyser Ltd. London, UK.

Canby, J.V. 2002. Falconry (Hawking) in Hittite Lands. Journal of Near Eastern Studies, Volume 61 (3): 161-201.

Capainolo, P. and C.A. Butler. 2010. How fast can a falcon dive?: Fascinating answers to questions about Birds of Prey. Rutgers University Press, New Jersey, USA. 248pp.

Carson, Rachael. 1962. Silent Spring. Houghton Mifflin Harcourt, Boston, USA.

Chitty, D. 1967. The natural selection of self-regulatory behaviour in animal populations. Proceedings of the Ecological Society of Australia 2:51-78.

Cleaves, F.W. 1982. The Secret History of the Mongols. Published by Harvard University Press, Cambridge, Mass., USA.

Clements, J. 2005. A brief history of The Vikings: the last pagans or the first Europeans? Running Press Bok Publishers. Philadelphia, PA, USA. 273pp.

Clum, N.J. and T.J. Cade. 1994. Gyrfalcon (Falco rusticolus). The Birds of North America, No. 114. A. Poole and F. Gill, Eds. Philadelphia: The Academy of Natural Sciences, Washington, D.C. The American Ornithologists Union.

Coelho, P. 2006. Like the Flowing River. Harper Collins, UK. 245pp.

Dalby, D. 1965. Lexicon of the Mediaeval German Hunt. Walter de Gruyter and Co. Berlin, Germany.

Dale, S.F. 1994. Indian Merchants and Eurasian Trade, 1600-1750. Cambridge University Press, Cambridge, UK.

Davis, P.K. 1999. 100 Decisive Battles: from ancient times to present. Oxford Univ. Press, Oxford, NY, USA.

de Grubernatis, A. 1872. Zoological Mythology, or the Legends of Animals, Vol. 2. Trubner and Company, Leipzig, Germany. Cited in Skuker 2015.

de Rachewiltz, I. 2004. The Secret History of the Mongols: A Mongolian Epic Chronical of the 13th Century. Translated by Igor De Rachewiltz, Published by Brill Inner Asian Library, Leiden. 1,347 pp.

Dekker, D. and R. Ydenberg. 2004. Raptor predation on wintering dunlins in relation to the tidal cycle. The Condor 106(2): 415-419.

Dementiev, G.P. 1960. Der Gerfalke. Die Neue Brehm – Buckeriei, no. 264. A. Ziemen Verlag, Wittenberg Lutherstadt, Germany. [English translation, 1967, Typescript by Foreign Languages Division, Department of State, Canada.]

Dobney, K. 2001. Ancient Falconry. (http://www. firstscience.com/SITE/articles/dobney.asp)

Dunne, Peter. 1995. The Wind Masters: the lives of North American birds of prey. Houghton Mifflin Company. Boston. 263pp.

Ellis, D.H., C.H. Ellis, G.W. Pendleton, A.V. Panteleev, I.V. Rebrova and Yu.M. Markin. 1992. Distribution and color variation of Gyrfalcons in Russia. Journal of Raptor Research 26: 81-88

Ellis, D.H., M. Wink, and V. Moseikin. 2008. The Altai Falcon: is it a gyrfalcon, a saker, a unique species, a mutation, a color morph or a figment of the imagination? Anuario de la Asociación Española de Centrería y Conservación de Aves Rapaces [AECCA] 2007:8-19.

Elton, C.S. 1924. Periodic Fluctuations in the Number of Animals: Their Causes and Effects. British Journal of Experimental Biology 2: 119-163.

Enderson, J.H. 2005. Peregrine Falcons: Stories of the Blue Meanie. University of Texas Press, Austin, Texas, USA.

Ericson, G.P. and T. Tyrberg. 2004. Early History of the Swedish Avifauna: a review of the subfossil record and early written sources. Alnquist and Wiksell Intl.

Evensen, E. 2007. Gods of Asgard. Saga Publishers International. 168pp.

Fagan, B. 2000. The Little Ice Age: How climate made history 1300-1850. Basic Books, Perseum Book Group, NY, NY.

Fagan, B. 2008. The Great Warming: Climate Change and the Rise and Fall of Civilizations. Bloomsburg Press, NY, NY.

Falkdalen, U., M. Hornell-Willebrand, T. Nygard, T. Bergstrom, G. Lind, A. Nordin, and B. Warensjo. 2011. Relations Between Willow Ptarmigan (Lagopus lagopus) Density and Gyrfalcon (Falco rusticolus) Breeding

performance in Sweden. Pages 171-176, Volume II, In R.T. Watson, T.J. Cade, M. Fuller, W.G. Hunt and E. Potapov (Eds.), Gyrfalcons and Ptarmigan in a Changing World, The Peregrine Fund, Boise State University, Idaho, USA, 1-3 February 2011.

Fitzherbert, T. 2006. Religious diversity under Ilkhanid Rule c 1300 as reflected in the freer Bal'ami. Koninklijke Brill NV, Leiden, The Netherlands.

Friedland, K. 1964. The Hanseatic League and Hanse Towns in the Early Penetration of the North. Arctic 37: pp. 539-543.

Fox, N. and E. Potapov. 2001. Altai Falcon: subspecies, hybrid or colour morph? Proceeding of 4th Eurasian Congress on Raptors, Seville, Spain, 25-29 September 2001.

Francis, C.S. 2011. The Lost Settlements of Greenland, 1342. Masters of Arts Thesis, California State University, Sacramento, California, 75pp.

Frisch, R. 1982. Birds of the Dempster Highway. Morriss Printing Company Ltd. Victoria, B.C. 98pp.

Fryer, John. 1689. A New Account of East India and Persia: being Nine Years Travels, 1678-1681. Edited by W. Crooke, Reprint: Millwood 1967. Cited in Allsen 2006.

Furmansky, D.Z. 2009. Rosalie Edge, hawk of mercy: the activist who saved nature from the conservationists. Univ. of Georgia Press, Athens, Georgia.

Gahbauer, M. 2002. Talon Tales. Canadian Peregrine Foundation. (http://www. peregrine foundation.ca/raptors/Gyrfalcon.html).

Gardarsson, A. 1971. Food ecology and spacing behavior of rock ptarmigan (Lagopus mutus) in Iceland. PhD Thesis, University of California, Berkeley.

Gesner, C. 1551. Historiae animalium. Volume 3. Christoffel Froschower, Zurich.

Ghazanfar, Ali Khan. 2015. Horses domesticated 4,000 years earlier than thought. Arab News, July 28, 2015.

Gilg, O., B. Sittler, and I. Hanski. 2009. Climate change and cyclic predator-prey population dynamics in the high Arctic. Global Change Biology, Volume 15, Issue 11, pages 2634-2652.

Gould, S. J. 1991. Bully for Brontosaurus: Reflections in Natural History. W.W. Norton and Company, New York, USA. 540 pp.

Grant, R.G. 2011. 1001 Battles that Changed the course of World History. Universe Publishing, New York, NY, USA.

Griffith, R.T.H. 1890. The Complete Rig Veda Translations. Classic Century Works, Kindle Edition. 2012.

Gruys, R.C. 1993. Autumn and Winter Movements and Sexual Segregation of Willow Ptarmigan. Arctic 46(3): 228-239.

Hagen, Y. 1952. The Gyr-Falcon (Falco R. Rusticolus L.) in Dovre Norway – some breeding records and food studies. Oslo. I Kommisjon Ros Jacob Dyrwad.

Hagerup, A.T. 1891. The Birds of Greenland. Library of the University of California – Biology request of Theodore S. Palmer. Translated by F.B. Arngrimson. Little, Brown and Co. Boston, Mass.

Haksoz, C. and D.D. Usar. 2011. Silk Road Supply Chains: a historical perspective. In Haksoz, C., S. Seshadri and A.V. Iyer (eds.). Managing Supply chains aon the Silk Road: strategy, performance and risk. Taylor and Francis Group, Boca Raton, Florida, USA.

Hannon, S.J., K. Martin, and J.O. Schieck. 1988. Timing of reproduction in two populations of willow ptarmigan in Northern Canada. The Auk 105:330-338.

Hardaswick, V. and K.L. Christopher. 2011. High Flying Gyrfalcons: The Guided Development Program. Western Sporting, Sheridan, Wyoming, USA.

Haskins, C.H. 1922. Science at the Court of the Emperor Frederick II. The American Historical Review 27(4): 669-694. Cited in Shahid 2009.

Hayes, R. 1977. The prey utilized by nesting peregrine falcons, gyrfalcons and golden eagles near the Dempster Highway Corridor. Yukon Wildlife Branch, internal publication, Whitehorse, Yukon Canada.

Hewitt, J.F.K. 1895. The ruling races of prehistoric times in India, South-western Asia and Southern Europe. Vol. II. Archibald Constable Company, Edinburgh, Scotland.

Hickey, J.J. and D.W. Anderson. 1968. Chlorinated hydrocarbons and eggshell changes in raptorial and fish-eating birds. Science 162: 271-273.

Hitti, P.K. 2002. History of the Arabs. 10th edition. Palgrave Macmillan Press, London, UK. Cited in Shahid 2009.

Hohn, E.O. 1969. Eskimo bird names at Chesterfield Inlet and Baker Lake, Keewatin, Northwest Territories. Arctic 22 (1): pp. 72-76.

Hopkins, P. 1980. Foreign Devils of the Silk Road: The Search for the Lost Treasure of Central Asia. John Murray Publishers, Great Britain.

Hudson, P.J., A.P. Dobson, and D. Newborn. 2003. Parasitic worms and population cycles of red grouse. http://www.personal.psu.edu/pjh18/downloads/119_Hudson_et_al_2003_Parasitic_worms_and_cycles_Berryman.pdf

Irving, L. 1953. The Naming of Birds by Nunamiut Eskimos. Arctic 6 (1): pp. 35-43.

Irving, L. 1958. Naming of Birds as part of the Intellectual Culture of Indians at Old Crow, Yukon Territory. Arctic 11 (2): pp 119-122.

Isaev, A.P. 2011. Changes in Ptarmigan Numbers in Yakutia. Pages 259-266, In R.T. Watson, T.J. Cade, M. Fuller, W.G. Hunt and E. Potapov (Eds.), Gyrfalcons and Ptarmigan in a Changing World, The Peregrine Fund, Boise State University, Idaho, USA, 1-3 February 2011.

Jacobsen, P.B. 1848. "Bidrag tile n Skildring af Falkevaesentet of Falkejagten forhen, havnlig I Danmark." In: Nyt Historisk Tidsskrift. Andet bind. Den Danske historiske forening, Kiobenhavn. Pp 307-414. Cited in Aegisson 2015.

Jenkins, M.A. 1982. Some behavioral aspects of gyrfalcon (Falco rusticolus) breeding biology. In: Ladd, W.N., and Schempf, P.F., eds. Raptor management and biology in Alaska and Western Canada. Proceedings of a Symposium and Workshop, 17 – 20 February 1981, Anchorage, Alaska. Anchorage: United States Fish and WildlifeService.205-216.

Jennbert, K. 2011. Animals and Humans: recurrent symbiosis in archaeology and Old Norse religion. Nordic Academic Press, and Kristina Jennbert, Lund, Sweden.

Johansen, K. and A. Ostlyngen. 2011. Ecology of the Gyrfalcon in Finnmark Based on Data from Two 11-Year Periods 150 Years Apart. Pages 141-160, In R.T. Watson, T.J. Cade, M. Fuller, W.G. Hunt and E. Potapov (Eds.), Gyrfalcons and Ptarmigan in a Changing World, The Peregrine Fund, Boise State University, Idaho, USA, 1-3 February 2011.

Johnson, J.A. and K. Burnham. 2012. Timing of breeding and offspring numbers co-vary with plumage color among gyrfalcons. Ibis 155(1): 177-188.

Johnson, O.W., L. Fielding, J.W. Fox, R.S. Gold, R.H. Goodwell, and P.M. Johnson. 2011. Tracking the Migrations of the Pacific Golden Plovers (Pluvialis fulva) between Hawaii and Alaska: New insight on flight performance, breeding ground destinations, and nesting from birds carrying light level geolocators. Wader Study Group Bull. 118(1): 26–31.

Jones, M.P, K.E. Pierce, D. Ward. 2007. Avian Vision: a review of form and function with special consideration of birds of prey. Journal of Exotic Pet Medicine. Vol. 16(2): 69-87. Cited in Walsh and Milner 2011.

Kahn, P. and F.W. Cleaves. 1999. The Secret History of the Mongols: the Origin of Chingis Khan. Cheng and Tsui Company, Boston, MA, USA.

Kalyakin, V.N. 1989. Birds of Prey in Ecosystems of the Extreme North. In Yu. I. Chernov, Yu.I. (ed.), Birds in the Natural Communities of the Tundra Zone, 51-112. Nauka, Moscow (in Russian). Cited in Potapov and Sale 2005.

Kishinsky, A.A. 1988. Ornithofauna of Northeast Asia. Nauka, Moscow (in Russian). Cited in Potapov and Sale 2005.

Kohli, M.S. 2003. Miracles of Ardaas: incredible adventure and survival. Indus Publishing Company, New Delhi.

Kolbas, J. 2006. The Mongols in Iran: Chingiz Khan to Uljayto, 1220-1309. Routledge Press, London, UK. 432pp.

Komaroff, L. 2006. Beyond the Legacy of Genghis Khan. Brill Academic Publishers. Leiden, Netherlands.

Koskimies, P. 1998. Gyrfalcon studies in Finland in 1997. Blafoten 5:2. Cited in Potapov and Sale 2005.

Koskimies, P. 2011. Conservation Biology of the Gyrfalcon in Northern Fennoscandia. Pages 95-124, In R.T. Watson, T.J. Cade, M. Fuller, W.G. Hunt and E. Potapov (Eds.), Gyrfalcons and Ptarmigan in a Changing World, The Peregrine Fund, Boise State University, Idaho, USA, 1-3 February 2011.

Kovalev, R.K. 2012. Russian history, Volume 39, Issue 4: page 460-517.

Lamb, H.H. 1966. The changing climate. Methuen, London, UK.

Lamb, H.H. 1995. Climate, History and the Modern World (2nd Edition). Routledge, London, UK. 433pp.

Lederer, R.J. 2016. Beaks, bones and bird songs: how the struggle for survival has shaped birds and their behavior. Timber Press, Portland, Oregon, USA. 288pp.

Lejeune, J. website http://www.falconscanada. com/Gyrfalcons.com.html

Leopold, Aldo. 1949. A Sand County Almanac, and sketches here and there. Oxford University Press. 81pp

Lefebvre, L. 2005. Bird IQ test takes flight. EurekAlert. http://www.eurekalert.org/pub_ releases/2005-02

Leviton, R. 2012. My Pal Blaise – notes on a 60 billion year friendship. Indiana University, Bloomington Indiana, USA.

Lind, G. and A. Nordin. 1995. Indum – ett productivt jaktfalk – revir. Faglar I Jamtlnd Harjedalen 2, 2-4. Cited in Potapov and Sale 2005.

Lindholm, D. and D. Licolle. 2007. The Scandinavian Baltic Crusades 1100-1500. Osprey Publishing, Oxford, UK.

Loehle, C. and J.H. McCulloch. (2008). Correction to: A 2000-Year Global Temperature Reconstruction Based on Non-Tree Ring Proxies. Energy and Environment 19(1), 93–100. http:// www.econ.ohio-state.edu/jhm/AGW/Loehle.

Mabie, H.W. 1882. Norse stories retold from the Eddas. Roberts Brothers, Boston.

Magnus, Albertus. ~1260. De falconibus, in De Animalibus. Cited in Oggins 2004. (http:// en.wikipedia.org/wiki/Albertus_Magnus).

Magnusson, S. (1785). "Lysing Gullbringu – og Kjosarsyslu." In: Landnam Ingolfs Safn til sogu bess. I. bindi. Felagio Ingolfur, Reykjavik. Pages 1-96. 1935-1936. Cited in Aegisson 2015.

Martin, K. and S. Wilson. 2011. Ptarmigan in North America: Influence of Life History and Environmental Conditions on Population Persistence. Pages 45-54 Volume I, In R.T. Watson, T.J. Cade, M. Fuller, W.G. Hunt and E. Potapov (Eds.), Gyrfalcons and Ptarmigan in a Changing World, The Peregrine Fund, Boise State University, Idaho, USA, 1-3 February 2011.

Matthiessen, P. 1994. The Wind Birds. Mariner Books, Houghton Mifflin Harcourt, New York. 168 pp.

McCaffery, B.J., T.L. Booms, T.C.J. Doolittle, F. Broerman, J.R. Morgart, and K.M. Sowl. 2011. The Ecology of Gyrfalcons (Falco rusticolus) on the Yukon-Kuskokwim Delta, Alaska. Pages 191-176 Volume 1, In R.T. Watson, T.J. Cade, M. Fuller, W.G. Hunt and E. Potapov (Eds.), Gyrfalcons and Ptarmigan in a Changing World, The Peregrine Fund, Boise State University, Idaho, USA, 1-3 February 2011.

McCarthy, D.P. and D.J. Smith. 1995. Growth Curves for Calcium Tolerant Lichens in the Canadian Rockies. Arctic and Alpine Research, vol. 27 (3), pp. 290-297.

McGhee, R. 2001. Ancient People of the Arctic. UBC Press, Vancouver, B.C., Canada. 244pp.

McIntyre, C.L., D.C. Douglas, and L.G. Adams. 2009. Movements of juvenile Gyrfalcons from Western and Interior Alaska following departure from their natal areas. J. of Raptor Research 43:99-109.

McIntyre, C.L., L.G. Adams, and R.E. Ambrose. 1994. Using satellite telemetry to monitor movements of Gyrfalcon in northern Alaska and the Russian Far East. Journal of Raptor Research 28:61.

McLaughlin, R. 2016. The Roman Empire and the Silk Routes: The ancient world economy and the empires of Parthia Central Asia and Han China. Pen and Sword Books, Ltd. South Yorkshire, UK.

Mechnikova, S, M. Romanov, and N. Kudryavtsev. 2011. Change in Numbers and Nesting Ecology of the Gyrfalcon in the Yamal Peninsula, Russia, from 1981-2010. Pages 205-212, In R.T. Watson, T.J. Cade, M. Fuller, W.G Hunt, and E. Potapov (Eds.), Gyrfalcons and Ptarmigan in a Changing World, The Peregrine Fund, Boise State University, Idaho, USA, 1-3 February 2011.

Mena, D. 2007. Adventure Guide to Hungary. Hunker Publishing Inc. Edison, New Jersey.

Morozov, V.V. 2011. Ecological Basis for the Distribution and Breeding of Gyrfalcons in the Tundra of European Russia and Preconditions for Spreading to New Grounds. Pages 229-238, In R.T. Watson, T.J. Cade, M. Fuller, W.G. Hunt and E. Potapov (Eds.), Gyrfalcons and Ptarmigan in a Changing World, The Peregrine Fund, Boise State University, Idaho, USA, 1-3 February 2011.

Mosbech, A. and S.R. Johnson. 1999. Late Winter Distribution and Abundance of Sea-Associated Birds in Southwest Greenland, the Davis Straight and Southern Baffin Bay. Polar Research 18(1):1-17.

Mossop, D.H. 2011. Long-term Studies of Willow Ptarmigan and Gyrfalcon in the Yukon Territory: A Collapsing 10-year Cycle and Its Apparent Effect on the Top Predator. Pages 323-336, In R.T. Watson, T.J. Cade, M. Fuller, W.G. Hunt and E. Potapov (Eds.), Gyrfalcons and Ptarmigan in a Changing World, The Peregrine Fund, Boise State University, Idaho, USA, 1-3 February 2011.

Mossop, D.H. and R.D. Hayes. 1994. Long-term trends in the breeding density and productivity of Gyrfalcon Falco rusticolu in the Yukon territory, Canada. In B-U. Meyburg and R. Chancellor (eds.), Raptor Conservation Today, 403-413. Pica Press, Robertsbridge.

Mowat, F. 1963. Never Cry Wolf. McClelland and Stewart. 176 pp.

Nelson, M. 1956. Observations of Gyrfalcon in training and hunting. Journal of the Falconry Club of America 2: 21-25. Cited in Potapov and Sale 2005.

Newton, I. 2011. Conference Summary. Pages 5-10, Volume I, In R.T. Watson, T.J. Cade, M. Fuller, W.G. Hunt and E. Potapov (Eds.), Gyrfalcons and Ptarmigan in a Changing World, The Peregrine Fund, Boise State University, Idaho, USA, 1-3 February 2011.

Nicolle, D. 1990. Attila and the Nomad Hordes. Osprey Publications – Reed International Books Ltd., Michelin House, London; 64pp.

Nicolle, D. and V. Shpakovsky. 2001. Genghiz Khan's Mongol Invasions of Russia. Osprey Publications – Reed International Books Ltd., Michelin House, London.

Nielsen, O.K. 1986. Population ecology of the Gyrfalcon in Iceland with comparative notes on the merlin and the raven. unpublished PhD Thesis. Cornell University.

Nielsen, O.K. 1991. Kynproskaaldur og atthagatryggd falka. Natturfrodistofnunar 60: 135-143. Cited in Potapov and Sale 2005.

Nielsen, O.K. 1999. Gyrfalcon predation on Ptarmigan: numerical and functional responses. Journal of Animal Ecology 68: 1034-1050.

Nielsen, O.K. 2011. Gyrfalcon population and reproduction in relation to rock ptarmigan numbers in Iceland. Volume II, pp 21-48, In R.T. Watson, T.J. Cade, M. Fuller, W.G. Hunt and E. Potapov (Eds.), Gyrfalcons and Ptarmigan in a Changing World, The Peregrine Fund, Boise State University, Idaho, USA, 1-3 February 2011.

Nielsen, O.K. and T.J. Cade. 2017. Gyrfalcon and ptarmigan predator-prey relationship. Pages 43-74 in D.L. Anderson, C.J.W. McClure, and A. Franke, editors. Applied raptor ecology: essentials from gyrfalcon research. The Peregrine Fund, Boise, Idaho, USA. (https://doi.org/10.4080/ore.2017/003)

Nielsen, O.K. and T.J. Cade. 1990. Annual cycle of the Gyrfalcon in Iceland. National Geographic Research 6: 41-62.

Nielsen, O. K., and G. Pétursson. 1995. Population fluctuations of gyrfalcon and ptarmigan: analysis of export figures from Iceland. Wildlife Biology 1:65-71.

Nikitin, A. 1475. Khozhenie za tri moria.2nd ed. Moscow-Leningrad, 1958. http:// tangentialia.wordpress.com/2009/06/22. Cited in Allsen 2006

Nygard, T., U. Falkdalen, and H. Engstrom. 2011. The dispersal of satellite-tagged juvenile gyrfalcons from an area of wind-farm development in the Swedish Mountains. Pages 161-170, In R.T. Watson, T.J. Cade, M. Fuller, W.G. Hunt and E. Potapov (Eds.), Gyrfalcons and Ptarmigan in a Changing World, The Peregrine Fund, Boise State University, Idaho, USA, 1-3 February 2011.

Nystrom, J., L. Dalen, P. Hellstrom, J. Ekenstedt, H. Angleby, and A. Angerbjorn. 2006. Ratio of rock to willow ptarmigan. Journal of Zoology 269:57-64.

Obst, J. 1994. Tree-nesting by the Gyrfalcon (Falco rusticolus) in the western Canadian Arctic. Journal of Raptor Research 28: 4-8.

O'chee, W. 2008. History of the Yuezhi. http:// www.chinahistoryforum.com/ topic/28467- history-of-the-yuezhi/

Oggins, R.S. 2004. The Kings and Their Hawks: Falconry in Medieval England. Yale University Press, New Haven, Connecticut. 251pp.

Palmer, R.S. 1988. Handbook of North American Birds, Vol. 5. Yale University Press, New Haven, Connecticut, USA and London, UK.

Patterson, R.L. 1879. Some of the wading birds frequenting the Belfast Lough. Pages 55-86 in Belfast Natural History and Philosophical Society 1878-79; published in 1880, London, UK.

Pedersen, A.O., M. Lier, H. Routii, H.H. Christensen, E. Fuglei. 2006. Co-feeding between Svalbard Rock Ptarmigan (Lagopus muta hyperborean) and Svalbard Reindeer (Rangifer tarandus platyrhynchus). Arctic 59(1): 61-64.

Pedersen, A.O., M.A. Blanchet, M. Hornell- Willeram, JU. Jepsen, M. Biuw, and E. Fuglei. 2014. Rock Ptarmigan (Lagopus muta) breeding habitat use in northern Sweden. Journal of Ornithology 155: 195-209.

Pepper, A. 2012. 10 words of phrases derived from falconry. Listverse http://listverse. com/2012/06/09/10-words-of-phrases-derived- from-falconry/.

Perry, M. 2013. Western Civilization: A brief history, tenth edition. Electronic version. Wadsworth, Boston, Mass. USA.

Phillott, D.C. 1908. Translation of The Baz- nama-yi Nasiri: A Persian Treatise on Falconry, written by Taymur Mirza in the 19th Century. Published by Bernard Quaritch, London.

Platt, J.B. 1977. Gyrfalcon courtship and early breeding behaviour on the Yukon north slope. Sociobiology 15: 43-69.

Platt, J.B. and C.E. Tull. 1977. A study of wintering and nesting gyrfalcons on the Yukon North Slope during 1975 with emphasis on their behavior during experimental overflights by helicopters. Arctic Gas Biol. Rep. Ser. 35(1): 1-90.

Pliny the Elder. 77. Pliny's Natural History. Translated by H. Rackham, W.H.S. Jones, and E.E. Eichholz, August 2014.

Pokrovsky, I. and N. Lecompte. 2011. Comparison of Gyrfalcon Distributions Between the Palearctic and Nearctic. Pages 161-170, In R.T. Watson, T.J. Cade, M. Fuller, W.G. Hunt and E. Potapov (Eds.), Gyrfalcons and Ptarmigan in a Changing World, The Peregrine Fund, Boise State University, Idaho, USA, 1-3 February 2011.

Polo, M. (1324). Travels of Marco Polo. Cited in the edition of Garden City Publishing, New York 1930.

Poole, K.G. and R.G. Bromley. 1988. Natural history of the Gyrfalcon in the Central Canadian Arctic. Arctic 41:31-38.

Poroarson, B. 1957. Islenzkir falkar. Safn til sogu islands og islenzkra bokmennta, annar flokkur. I.5, 168pp, cited in Nielsen and Petursson 1995.

Potapov, E. 2011a. Gyrfalcon Diet: Spatial and Temporal Variation. Pages 55-64, Volume I, In R.T. Watson, T.J. Cade, M. Fuller, W.G. Hunt and E. Potapov (Eds.), Gyrfalcons and Ptarmigan in a Changing World, The Peregrine Fund, Boise State University, Idaho, USA, 1-3 February 2011.

Potapov, E. 2011b. Gyrfalcons in Russia: Current Status and Conservation Problems. Pages 191-196, Volume II, In R.T. Watson, T.J. Cade, M. Fuller, W.G. Hunt and E. Potapov (Eds.), Gyrfalcons and Ptarmigan in a Changing World, The Peregrine Fund, Boise State University, Idaho, USA, 1-3 February 2011.

Potapov, E. and R. Sale. 2005. The Gyrfalcon. T&AD Poyser, UK.

Ratcliffe, D. 2005. Lapland: A natural history. T&AD Poyser, UK.

Ratcliffe, D.A. 1967. Decrease in eggshell weight in certain birds of prey. Nature 215, 208-210. London.

Robinson, 2016. Gyrfalcon diet during the brood rearing period on the Seward Peninsula, Alaska, in the context of a changing world. MSc. Thesis, Boise State University, USA, August 2016.

Robinson, B.W., N. Paprocki, D.L. Anderson, and M.J. Bechard. 2017. First record of nestling relocation by adult gyrfalcon (Falco rusticolus) following nest collapse. The Wilson Journal of Ornithology, 129(1): pp. 216-221.

Rohner, C. 1997. Non-territorial floaters in Great Horned Owls: space use during a cyclic peak of snowshoe hares. Animal Behaviour 53:901-912.

Salby, M.L. 2012. Physics of the Atmosphere and Climate. Cambridge University Press, NY, NY, USA.

Salomonsen, F. 1950-1951. Gronlands fugle. Volume 1-3. Ejnar Munksgaard. Kobenhavn.

Salvin, F.H. and W. Brodrick. 1855. Falconry in the British Isles. Reprinted 1997, Beech Publishing House, Midhurst, UK. Cited in Shrubb 2013.

Sanchez, G.H. 1993. The ecology of wintering Gyrfalcons Falco Rusticolus in central South Dakota. MS thesis, Boise State University, Idaho.

Sandor, F. 2013. Magyar Origins – a 21st Century look at the origins of ancient Hungarians. Self-published.

Sankaran, A. 2012. How is hawk vision better than human vision anatomically (eye components and visual cortex)? http://www. quora.com.

Sarangerel. 2000. Riding Windhorses: A journey into the heart of Mongolian Shamanism. Destiny Books, Rochester, Vermont. 210 pp.

Schmidt, N.M., R. A. Ims, T.T. Hoye, O. Gilg, L.H. Hansen, J. Hansen, M.C. Forchhammer, and B. Sittler. (2012). Response of an arctic predator guild to collapsing lemming. The Royal Society, Sept. 12, 2012.

Seaver, K. 2010. The Epic Story of the Great Norse Voyages: the last Vikings. I.B. Tauris and Co. Ltd. NY, NY, USA.

Seaver, K. 1996. The Frozen Echo. Stanford University Press, Stanford, USA.

Secinski, W. 2006. The mystery of Genghis Khan: a historical novel, book 1. Infinity Publishing company, West Conshohocken, PA., USA.

Selas, V. and J.A. Kalas. 2007. Territory occupancy rate of goshawk and gyrfalcon: no evidence of delayed numerical response to grouse numbers. Oecologica 153(3): 555-561.

Service, R. 1973. The Collected Poems of Robert Service. Dodd, Mead & Company, New York.

Sethi, M.L. 2012. Mountainfit: Fjallsommar, Fjallsjalv. Self published, Goodreads.com. Also cited in Zelnio, K. 2012. http:// blogs.scientificamerican.com/WSS/post. php?blog=20&post=602

Shahid, I. 2009. Byzantium and the Arabs in the Sixth Century. Volume 2, Part 2. Dumbarton Oaks, Trustees for Harvard University.

Shailor, B.A. 1991. The Medieval Book. Univ. of Toronto Press, Toronto, Ontario.

Shergalin, J. 2011. Brief Review of Russian- language Literature on the Gyrfalcon. Pages 239-342, In R.T. Watson, T.J. Cade, M. Fuller, W.G. Hunt and E. Potapov (Eds.), Gyrfalcons and Ptarmigan in a Changing World, The Peregrine Fund, Boise State University, Idaho, USA, 1-3 February 2011.

Shrubb, M. 2013. Feasting, Fowling and Feathers: a history of exploitation of wild birds. T & AD Poyser, London, UK.

Sielicki, J. 2009. Falconry in Poland: an historical outline. Int. Journal of Falconry. Spring 2009, pages 47-48.

Skuker, K. 2015. Tracking the Turul – Hungary's Premier Mythological Bird. Self-published, Goodreads.com.

Smith, S. 2013. Desert finds challenge horse taming ideas. BBC News, February 26, 2013.

Spielvogel, J.J. 2008. Western Civilization: a brief history. Clark Baxter, Thompson Learning Inc, Thomason Wadsworth.

Stevens, R. 1956. The Gyrfalcon. Journal of the Falconry Club of America 1:18-22, cited in Potapov and Sale 2005.

Steves, R. and C. Hewitt. 2013. Rick Steves' Budapest. Google ebook, Avalon Travel.

Streeter, Stephanie. 2001. The Eyes Have It (Raptor eyes, that is). From Delaware Valley Raptor Centre Journal, Fall/Winter 2001.

Taknet, D.K. 2013. Jaipur: Gem of India. Published by IIME/DK Taknet.

Taylor, T.G. 1970. How an Eggshell is Made. Scientific American, March, pages 88-95.

Topsell, E. circa 1658. The Fowles of Heauen or History of Birds. In T.P. Harrison and F.D. Hoeniger (eds). University of Texas, Austin, Texas. 1972.

Twitchett, D. and K.P. Tietze. 1994. Alien Regimes and Border States 907-1368. In Franke, H. and D. Twitchett (eds.). The Cambridge History of China, Volume 6. University of Cambridge, Cambridge, UK.

Tyrberg, T. 2002. The Archaeological record of domesticated and farmed birds in Sweden. Acta zoologica cracoviensia 45: 215-231.

Vajda, E. 2012. Geography, demography and time depth: Explaining how Dene-Yeniseian is possible. Dene-Yeniseian Workshop. March 24, 2012, Alaska Native Language Workshop, University of Alaska, Fairbanks, Alaska.

Van de Wall, J.W.M. 2004. The Loo Falconry: the Royal Loo Hawking Club 1839. Hancock House Publishers, Victoria, B.C., Canada. Cited in Shrubbs 2013.

Van Geel, B., N.A. Bokovenko, N.D. Burova, K.V. Chuqunov, V.A. Dergachev, and V.G. Dirksen. 2005. Climate Change and the sudden expansion of the Scythian culture after 850 B.C. Publication of the Universiteit van Amsterdam, Russian Academy of Sciences. (http://dare.uva.nl/ record/170200)

Vaughan, R. 1992. In Search of Arctic Birds. T and AD Poyser Ltd., London, UK.

Wallis, R.J. 2017. 'As the Falcon Her Bells' at Sutton Hoo? Falconry in the Early Anglo- Saxon England. Archaeological Journal (online publication), April 10th, 2017.

Walsh, S. and A. Milner. 2011. Evolution o the Avian Brain and Senses. Chapter 11, in G. Dyke and G. Kaiser (eds.), Living dinosaurs: The Evolutionary History of Modern Birds. John Wiley and Sons, West Sussex, UK.

Watson, A., R. Moss, P. Rothery. 2000. Weather and synchrony in 10 year population cycles of rock ptarmigan and red grouse in Scotland. Ecology 81: 2126-2135.

Watson, A., R. Moss, S. Rae. 1998. Population dynamics of Scottish cock ptarmigan cycles. Ecology 79: 1174-1192.

Wells, H.G. 1922. A Short History of the World. Cassell and Co. Ltd., UK.

White, R.C.R. 2011. Changes in cyclic availability of food. Basic and applied Ecology, Vol. 12(6): 481-487.

White, C.M. and T.J. Cade. 1971. Cliff-nesting raptors and ravens along Colville River in Arctic Alaska. Living Birds 10: 107-150.

White, C. and R.B. Weeden. 1966. Hunting methods of gyrfalcons and behavior of their prey. Condor 68(5): 517-519.

Williams, B.F. (ed.) 1996. Woman Out of Place: the gender of agency and the race of nationality. Routledge, N.Y., N.Y. USA.

Wingfield, J.C. and M. Ramenofsky 2011. In advances and the study of behaviour, Vol 43. Academic Press, London, UK.

Wingmasters 2001. http://www.wingmasters.net

Wink, M., H. Sauer-Gurth, D. Ellis, R. Kenward. 2004. Phylogenetic relationships in the Hierofalco complex (Saker – Falco Cherrug, Gyrfalcon – F. rusticolus, Lanner – F. biarmicus, Laggar Falons). In Chancellor, R.D and B. Meyburg (eds.). Raptor Worldwide. WWGBP/ MME, pp. 499-504.

Wittfogal, K.A. and C. Feng. 1949. History of Chinese Society: Liao (907-1125). Transactions of the American Philosophical Society, Vol. 36. 1946.

Wood, F. 2002. The Silk Road: 2,000 Years in the Heart of Asia. University of California Press, Berkeley and Los Angeles, California, USA

Wood, C. and F. Fyfe. 1943. The Art of Falconry. "De Arte Venandi cum Avibus of Frederick II of Hohenstaufen". Stanford University Press, Stanford, California.

Xiaodi, Y., R. Chang, and N. Fox. 2001. The history of falconry in China. Falco (Newsletter of the Middle East Falconry Research Group): Issue 20, page 13.

Ydenberg, R.C., D. Dekker, G. Kaiser, P.C.F. Shepherd, L.E. Ogden, K. Rickards and D.B. Lank. 2010. Winter body mass and over-ocean flocking as components of danger management by Pacific dunlins. BMC Ecol. 10:1.

NORMAN BARICHELLO

NORMAN BARICHELLO IS A NORTHERN ECOLOGIST. HE has spent over 44 years in Canada's arctic and subarctic, as a biologist, naturalist guide and advisor to the Kaska Dena. As a biologist, Norman has participated in studies on polar bears, grizzly bears, wolves, caribou, elk, black-tailed deer, and birds of prey. His passion for gyrfalcons took hold during his academic studies at the University of British Columbia when he explored the relationship between gyrfalcons and ptarmigan in the Ogilvie Mountains, and has flourished over the last 35 years observing gyrfalcons in the Mackenzie Mountains. Norman currently lives in Whitehorse, and continues to spend his summers in the Mackenzie Mountains watching gyrfalcons.

CPSIA information can be obtained
at www.ICGtesting.com
Printed in the USA
LVHW071854091020
668326LV00030B/2319